신비로운 세계의
일러스트 모음집

하랑

출판사 : 하랑출판
주　　소 : 서울시 중구 퇴계로28길 8
전　　화 : 02-2263-3337

신비로운 세계의 일러스트 모음집

발행일 : 2025년 9 월 05일
출판사 : 하랑출판
주　　소 : 서울시 중구 퇴계로28길 8
전　　화 : 02-2263-3337

CONTENTS

WWW.BEHANCE.NET/PONTODEFUGA
TRANSFORMAMOS
OS SONHOS DE UM
RASCUNHO EM
ILUSTRAÇÕES
DE VERDADE.
PONTO DE FUGA
ILUSTRAÇÕES

WWW.BEHANCE.NET/PONTODEFUGA
TRANSFORMAMOS
OS SONHOS DE UM
RASCUNHO EM
ILUSTRAÇÕES
DE VERDADE.
PONTO DE FUGA
ILUSTRAÇÕES

GRAPHIC Eduardo Tavares   CORPORATION Exclam, Curitiba, Brazil   COPY Marcelo Russo

GRAPHIC Jaime Duque, Jhon Mondragon   CORPORATION Y&R, Bogotá   COPY Rafael Barthaburu, Juan Valdivieso

BAKE.
TOAST.
SPREAD.
GRIN.
REPEAT.
LOVE FOOD WITH LURPAK
LURPAK
SPREADABLE

OPEN A
WINDOW,
LET'S
MAKE THE
STREET
HUNGRY
LOVE FOOD WITH LURPAK
LURPAK
BUTTER

PHOTO Jonathan Lovekin   GRAPHIC David Goss   CORPORATION Saatchi & Saatchi, London

GRAPHIC Abe Garcia, Almendra Tijerina   CORPORATION Bromley Communications, San Antonio, Texas

GRAPHIC Ali Bati, Guney Soykan   CORPORATION DDB&Co., Istanbul   COPY Gul Kanik

Olaylar ve arka planları
CNN

Olaylar ve arka planları
CNN

Row, ROW, ROW your boat gently down The Stream,
till you get to THE RAPiDS &
MAke your-GiRLfriend
ScReaM!
BRITANNIA
bourbon
THE ORIGINAL
sugary sweet biscuit outside
dark creamy chocolate inside

YUM YUM MUMS
Delicious Choc Chip flavour is now available in a delightful new look Yum Yum pack. As yummy as you are, Mum.
"YUM YUM" choc chip

PHOTO Michael Meyersfeld   GRAPHIC Yvonne Hall   CORPORATION McCann Erickson, Johannesburg   COPY Anthea Weldon

FISH: NOW WITH CHILI.

FISH: NOW WITH CHILI.

PHOTO Carioca   GRAPHIC Raluca Stroe   CORPORATION BBDO, Moscow

PHOTO Nadav Kander   GRAPHIC Frank Dattalo, Isabela Ferreira

KEEP
LAST NIGHT
TO YOURSELF
FUN

KEEP
LAST NIGHT
TO YOURSELF
FUN

PHOTO Hans Starck   GRAPHIC Gabriel Mattar, Johannes Hicks, Stefan Schulte   CORPORATION DDB, Berlin

14999

USA
BASKETBALL

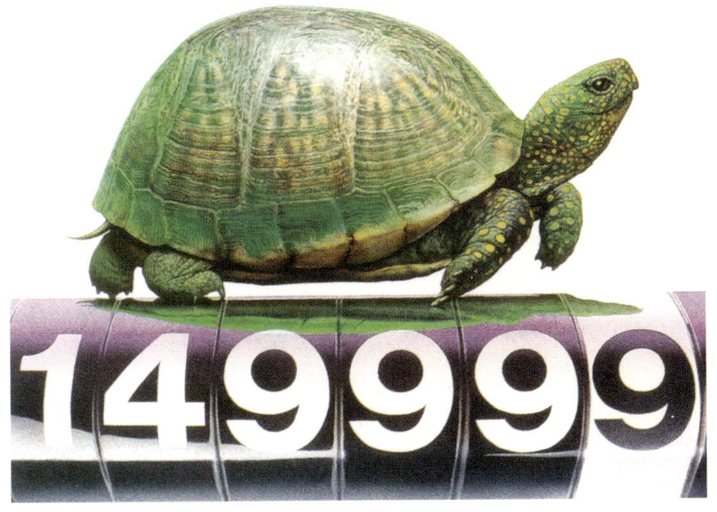

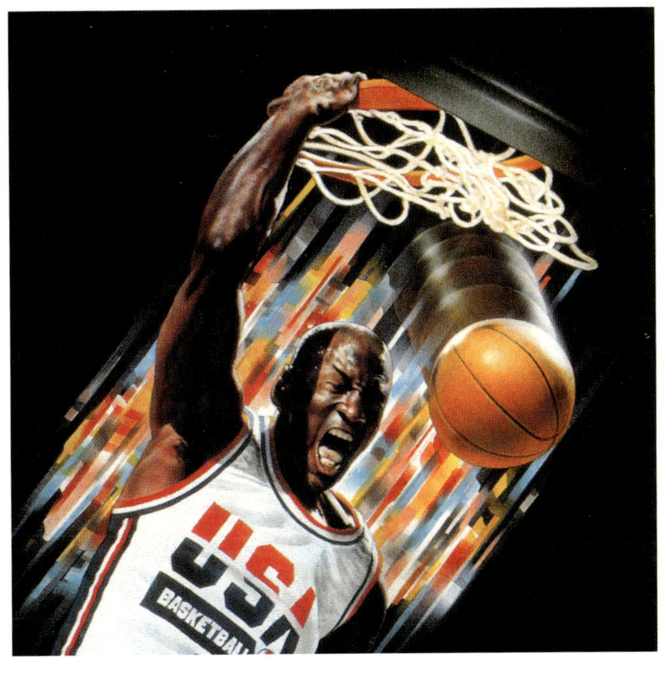

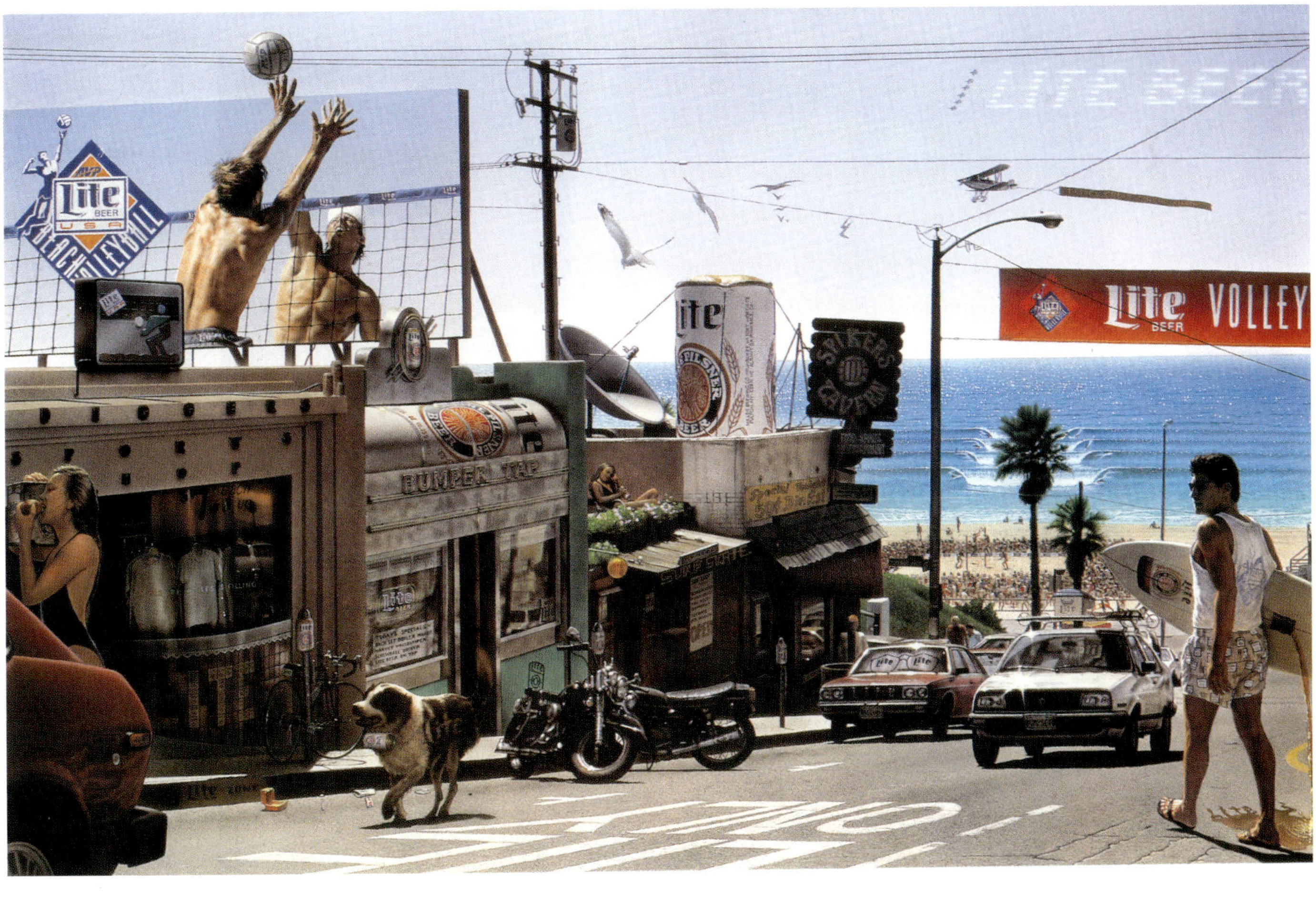
LITE BEER
Lite BEER VOLLEY
BUMPER TAP

CURES CAR SMELLS.
Ambi Pur

SEE AMERICA IN AN AIRSTREAM
AIRSTREAM

PHOTO Gary Sheppard   GRAPHIC Jay Furby   CORPORATION Jay Grey, Sydney   COPY David White

The weird and dreadful passion
of Morgana, who had no friends.
The first italian magazine of dark, gothic, electro, industrial music and culture
:Ritual:
YOUR DARK SIDE

The bizarre case of Vladimir,
who waketh up at midnight.
The first italian magazine of dark, gothic, electro, industrial music and culture
:Ritual:
YOUR DARK SIDE

PHOTO Winkler + Noah   GRAPHIC Chiara Catalani, Elisa Pazi   CORPORATION Ogilvy & Mather, Rome   COPY Elisa Pazi

挡得住!
碧浪

挡得住!
碧浪

GRAPHIC Marcus Cross   CORPORATION Cutwater, San Francisco

GRAPHIC Leopold Billard   CORPORATION Cutwater, San Francisco

The Eddy
December

PHOTO Davide Bodini   GRAPHIC Gabriele di Donato

PHOTO Ferdinando Galletti GRAPHIC Ferdinando Galletti CORPORATION Lowe Pirella, Milan COPY Atonino Munafò

LINE 1: 1-900-HOT-TALK
LINE 2: 1-900-544-COED
LINE 1: 1-900-DESIRES
LINE 2: 1-900-FAN-TASY
It's worth a second job.
The 2009 Titus El Guapo. MSRP $3370. To find out more about the new El Guapo™ and part time employment in your area visit titusti.com.
TITUS

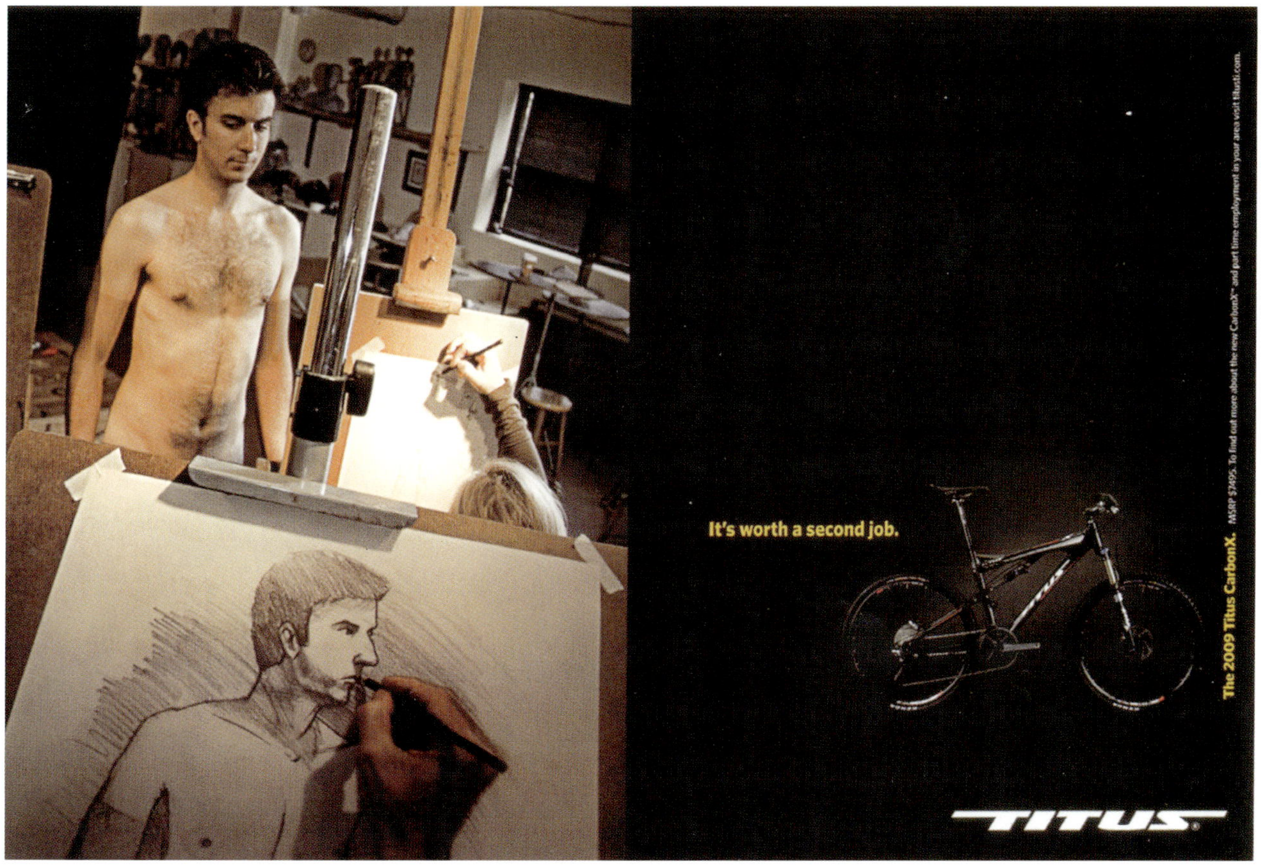

It's worth a second job.
The 2009 Titus CarbonX. MSRP $3495. To find out more about the new CarbonX™ and part time employment in your area visit titusti.com.
TITUS

PHOTO Andrew Chapman   GRAPHIC Thomas Dooley, Matt Leavitt   CORPORATION TDA Advertising & Design, Boulder

Alpina
NIKOTIN-
SPERRE
Covers nicotine perfectly.

Equipped with HyClean 4.5L extra large dustbag.
Miele

PHOTO Simone Rosenberg, Yona Heckel   GRAPHIC Nils Hipp, Tobias Rabe   CORPORATION Scholz&Friends, Hamburg

ECOMIST
eliminates odours

eliminates odours
ECOMIST

Michael McGovern

FLOWER
POWER YOUR
ROOM!

PHOTO Attila Hartwig   GRAPHIC Nicole Grözinger   CORPORATION Jung von Matt,

PHOTO Quah Kwan Guan, David Lok, Simon Ong

WWW.MOTHERSPRIDEONLINE.COM OR CALL 9310610011/12
Taekwondo.
Part of the curriculum.
MOTHER'S PRIDE
LOVE BLOSSOMS HERE

COPY Viral Pandya, Sabu Paul, Guneet Pandya

Fancy dress parties.
Part of the curriculum.
MOTHER'S PRIDE
LOVE BLOSSOMS HERE
KRISHNA KANH
THE MAN OF STEEL
WWW.MOTHERSPRIDEONLINE.COM OR CALL 9310610011/12

GRAPHIC Viral Pandya, Guneet Pandya   CORPORATION Out of the box, New Delhi   COPY Viral Pandya, Sabu Paul

Dance and music.
Part of the curriculum.
MOTHER'S PRIDE
LOVE BLOSSOMS HERE
WWW.MOTHERSPRIDEONLINE.COM OR CALL 9310610011/12

WWW.MOTHERSPRIDEONLINE.COM OR CALL 9310610011/12
Computers.
Part of the curriculum.
MOTHER'S PRIDE
LOVE BLOSSOMS HERE
GAME OVER

PHOTO Fernando Ziviani   GRAPHIC Gean Santos   CORPORATION Wow!, Curitiba, Brazil

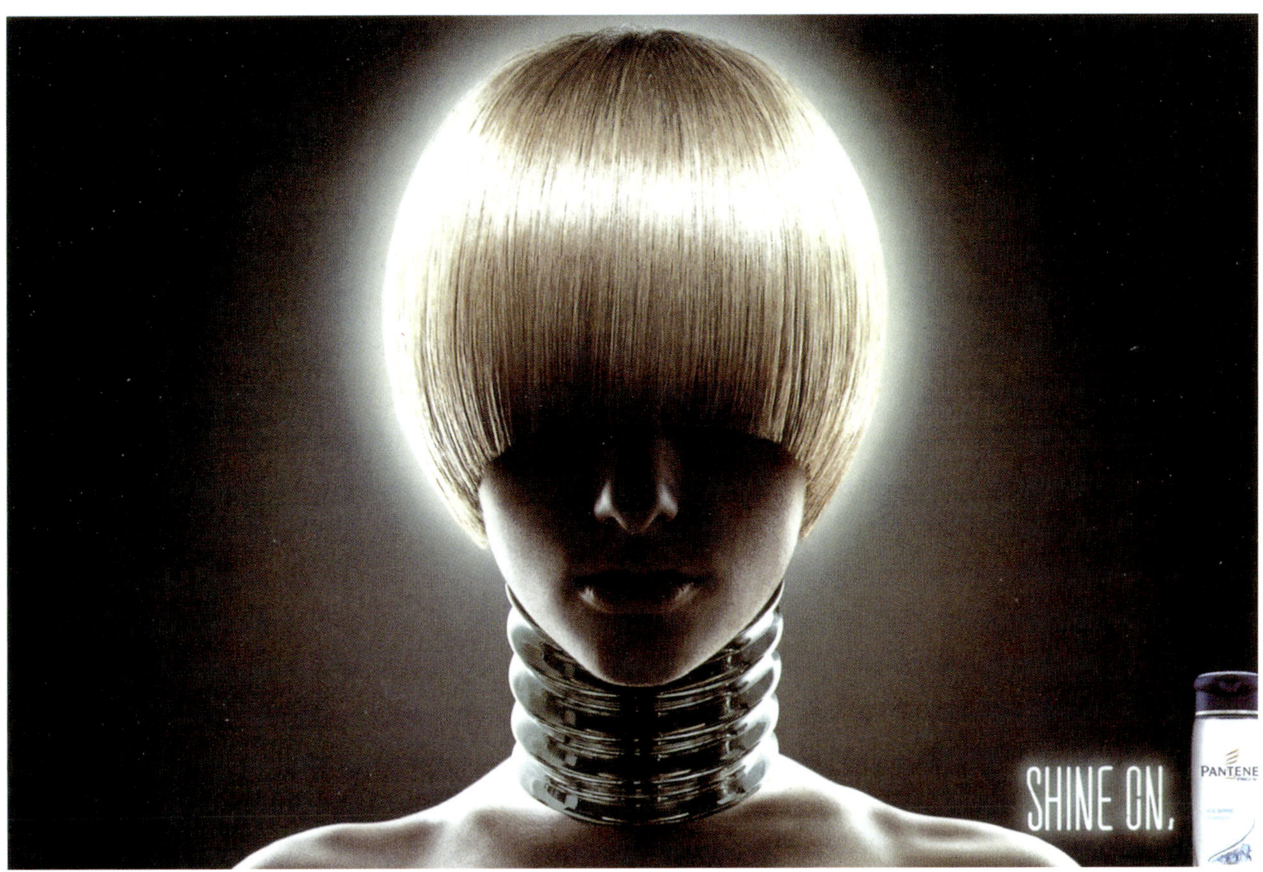

PHOTO Connie Hong   GRAPHIC Doris Leung   CORPORATION Grey, Hong Kong

BANG
www.daudio.com.br
Perfect union of sound and image.
D'audio

KABUM
www.daudio.com.br
Perfect sintony of sound and image.
D'audio

GRAPHIC Marcos Paulo Rodrigues de Carvalho   CORPORATION Tif Comunicação, Curitiba, Brazil   COPY Eduardo Visinoni

THESE are your TEXTBOOKS.
go PLACES.
UNIVERSAL TECHNICAL INSTITUTE

More office cubicles should
LOOK LIKE THIS.
go PLACES.
UNIVERSAL TECHNICAL INSTITUTE

REMOVE THE
red pylons from
YOUR FUTURE
MINI COOPER S
go PLACES.
UNIVERSAL TECHNICAL INSTITUTE

PHOTO Suresh Natarajan   GRAPHIC Ryan Menezes   CORPORATION McCann Erickson, Mumbai   COPY Ryan Menezes

PHOTO Arata Dodo GRAPHIC Kotaro Hirano

R DiGianni

FOR ONCE, WE SUPPORT THOSE WHO DON'T SAVE ENERGY.
ENEL WELCOMES YOU TO THE WORLD SWIMMING CHAMPIONSHIPS, ROME09.
Enel
ENERGY IN TUNE WITH YOU

PHOTO Paolo Gripo, Marlon Balangon   GRAPHIC Joel Limchoc, Dale Lopez

give us a call before it's too late. +90 216 316 26 18
rent a car to everywhere.
YEDiTEPE
RENT A CAR

give us a call before it's too late. +90 216 316 26 18
rent a car to everywhere.
YEDiTEPE
RENT A CAR

GRAPHIC Ali Altan Kelleci   CORPORATION Safari, Istanbul   COPY Y&R, Singapore

Objective   To contribute customer service, event
             and design experience in a marketing capacity

Education   Jacksonville University, Jacksonville, FL
             Bachelor of Business Administration in Marketing, May 2008
             Cumulative GPA: 3.25

             Lindenhurst High School, Lindenhurst, NY
             Cumulative GPA: 3.5

Experience   Sending out resumes. Waiting for the phone to ring.
             Checking to make sure the cord's plugged in.

Professional Team player who strives for productive collaboration.
Profile   Creative thinker with knowledge of marketing concepts.
             Self-starter who can work independently.
             Ability to handle multiple priorities and deadlines.

Skills   Proficient in word processing and spreadsheet analysis

Interests   Pottery, Reading, Triathlete

References Available Upon Request

GraduateCareerCoaching
graduatecareercoaching.com

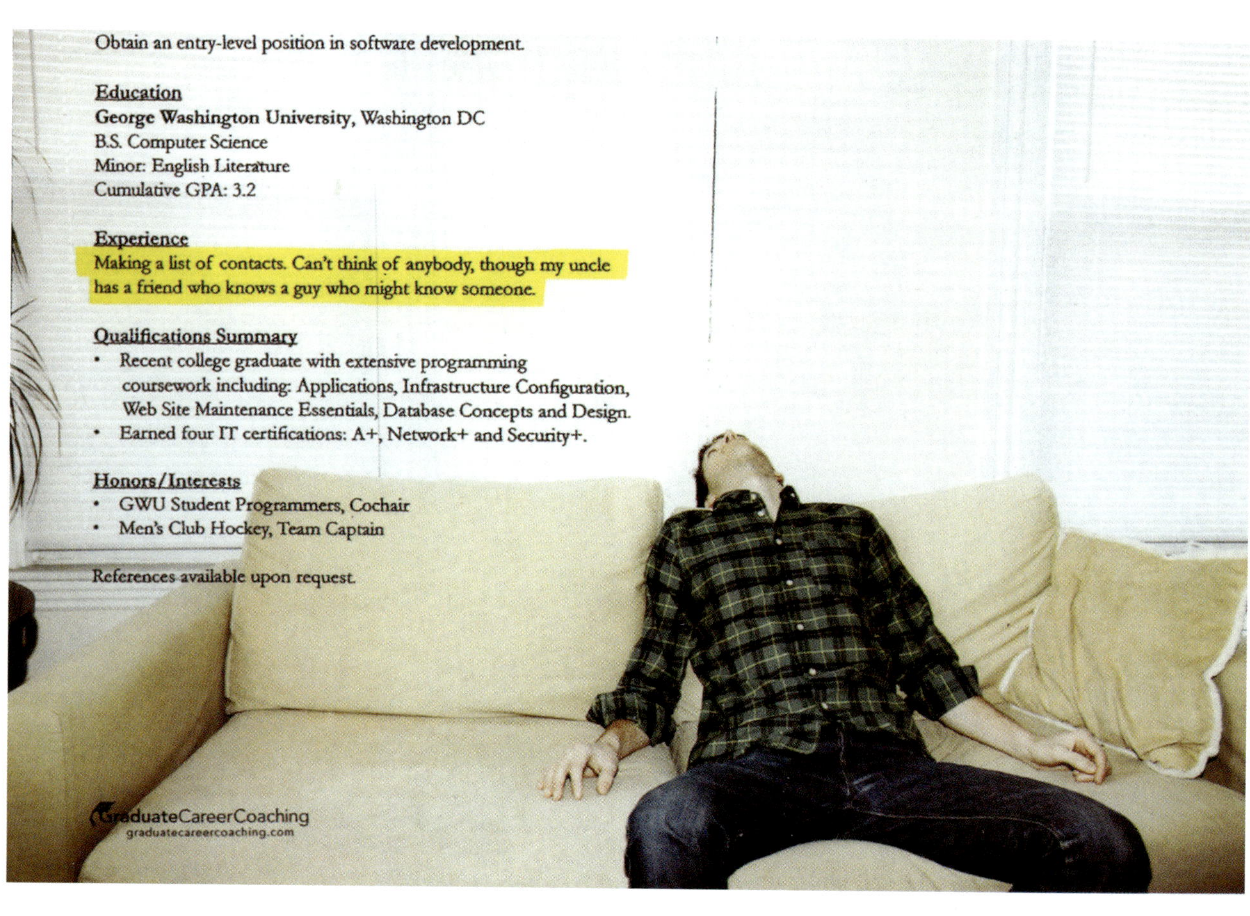

Obtain an entry-level position in software development.

Education
George Washington University, Washington DC
B.S. Computer Science
Minor: English Literature
Cumulative GPA: 3.2

Experience
Making a list of contacts. Can't think of anybody, though my uncle
has a friend who knows a guy who might know someone.

Qualifications Summary
• Recent college graduate with extensive programming
  coursework including: Applications, Infrastructure Configuration,
  Web Site Maintenance Essentials, Database Concepts and Design.
• Earned four IT certifications: A+, Network+ and Security+.

Honors/Interests
• GWU Student Programmers, Cochair
• Men's Club Hockey, Team Captain

References available upon request.

GraduateCareerCoaching
graduatecareercoaching.com

mentos rainbow
LACKING COLOR?
NOW YOU
CAN HAVE 7.

PLAYING
FOR
MENTOS
RAINBOW
mentos rainbow
LACKING COLOR?
NOW YOU
CAN HAVE 7.

PHOTO Jonathan Lovekin   **GRAPHIC** David Goss   **CORPORATION** Saatchi & Saatchi, London

PHOTO Ricardo Salaman GRAPHIC Miguel Angel Cerdeira, Patricio Céspedes CORPORATION Grey, Santiago

PENHALIGON'S
MERCHANTS of ATTRACTION
since 1870
Orange Blossom
Should it meet the nostrils of a gentleman he will be eyeing one's ankles forthwith.

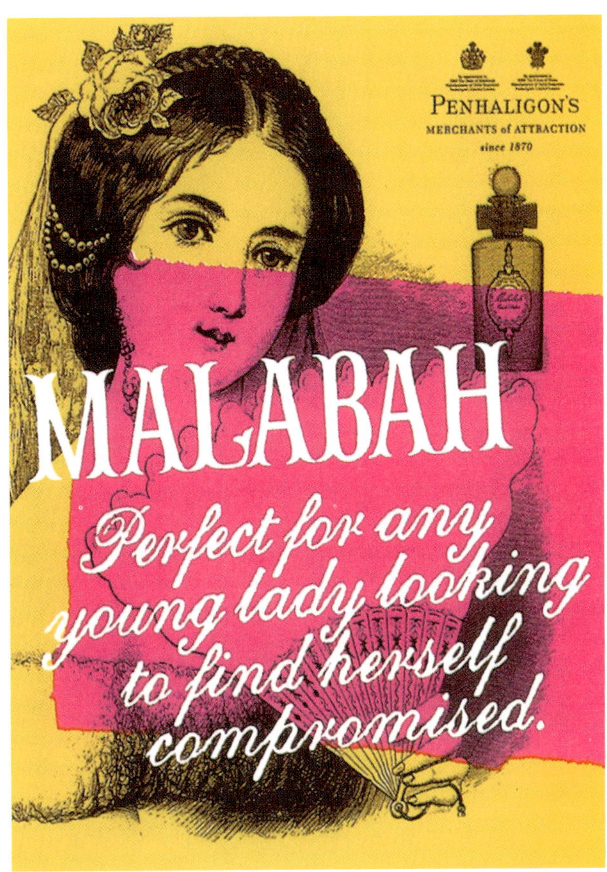

PENHALIGON'S
MERCHANTS of ATTRACTION
since 1870
MALABAH
Perfect for any young lady looking to find herself compromised.

PENHALIGON'S
MERCHANTS of ATTRACTION
since 1870
ENDYMION
A splash under one's top hat will aid all matters relating to the removal of a lady's corset.

PENHALIGON'S
MERCHANTS of ATTRACTION
since 1870
BLENHEIM BOUQUET
Gentlemen, apply and prepare for behaviour unbecoming of a lady.

GRAPHIC Christopher Bowsher   CORPORATION Dye Holloway Murray, London   COPY Frances Leach

104
R. A. F.
PATROL—
THE NAZI
DREAD

Findus
Fraich'Frites
Spécial friteuse
GRANNY'S FRIES
WITHOUT
THE GRANNY

Findus
Fraich'Frites
Spécial friteuse
GRANNY'S FRIES
WITHOUT
THE GRANNY

PHOTO Achim Lippoth   GRAPHIC Cedric Auzannet   CORPORATION Grey, Paris   COPY Benjamin Dessa

PHOTO Michael and Nic.com   **GRAPHIC** Yvonne Hall, Sarita Immelman   **CORPORATION** McCann Erickson, Johannesburg

PHOTO Natalie Shau   GRAPHIC Ian Broekhuizen

PHOTO Roberto Badin   GRAPHIC Sophie Guyon   CORPORATION TBWA, Paris   COPY Thomas Mayer

GRAPHIC Dominique Magnusson, Martin Friedlin   CORPORATION TBWA, Zurich   COPY Michael Kathe

THE IMPORTANT THING IS TO BE HEALTHY
SUCRILHOS

THE IMPORTANT THING IS TO BE HEALTHY
SUCRILHOS

PHOTO Edu Lopes   GRAPHIC Breno Balbino   CORPORATION Leo Burnett, São Paulo   COPY Fabio Nagano

10 11 12 1 2 3 4 5 6 7 8 9
Form 1040
Label
Use IRS label.
Otherwise.
please print
or type
LABEL HERE

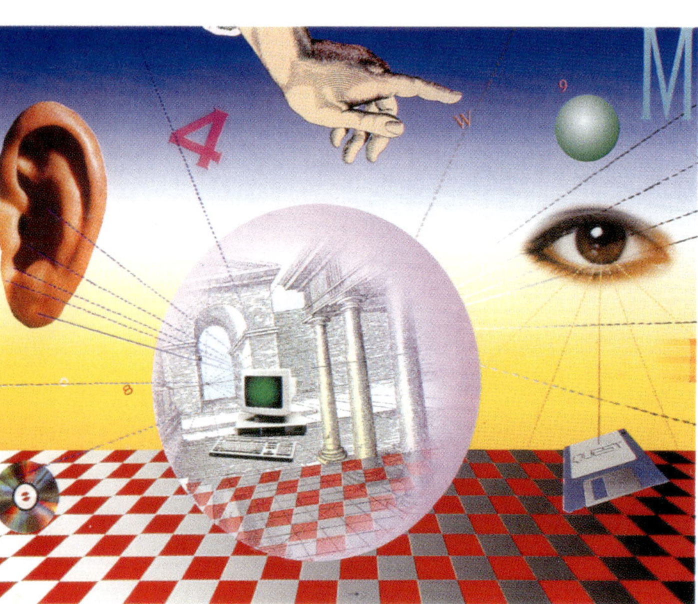
M
4
W
9
QUEST

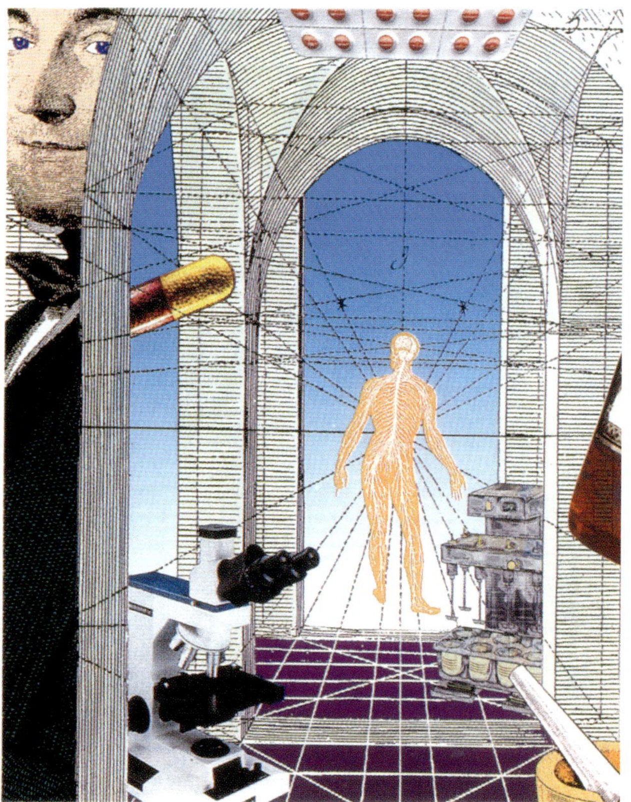

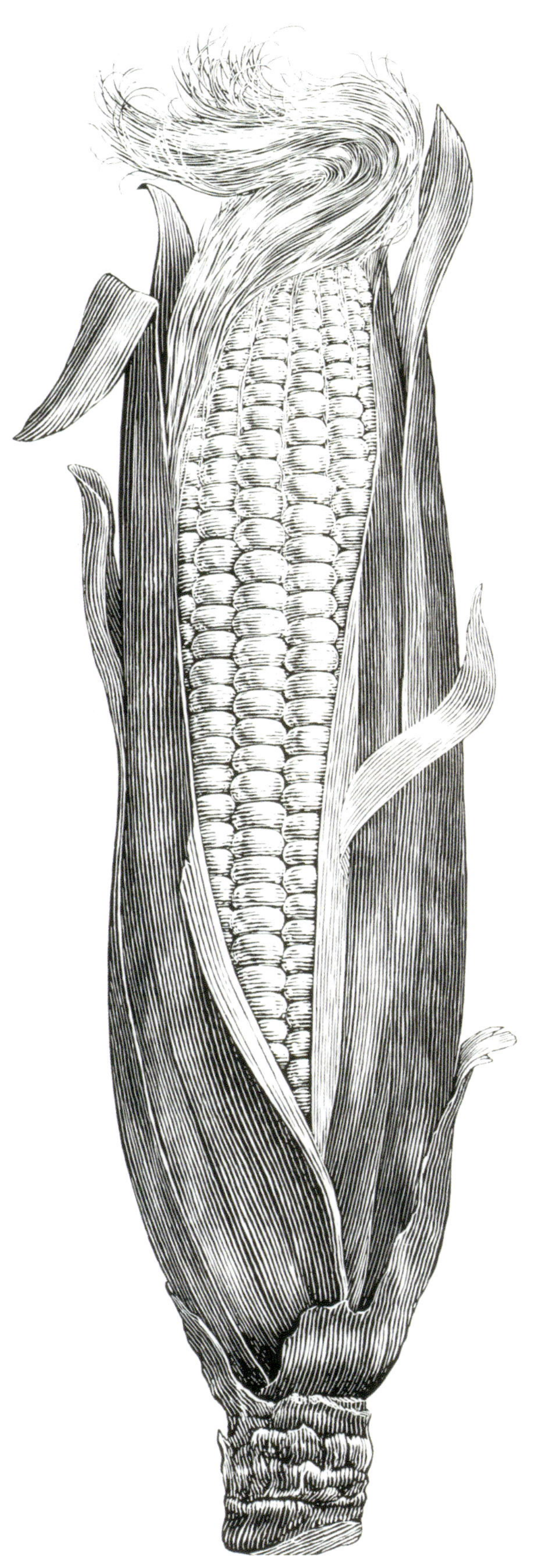

Fresh Step
Cats everywhere are having a hard time smelling their litter boxes.
freshstep.com

Fresh Step
Cats everywhere are having a hard time smelling their litter boxes.
freshstep.com

Long lasting glue.
Pattex

PHOTO Ciril Jazbec   GRAPHIC Saso Dornik   CORPORATION Imelda Ogilvy, Ljubljana, Slovenia   COPY Andrej Basa, Jure Pozun

shape up
your heart
Prevent cardiovascular diseases
using Salat Olive Oil
Salat

shape up
your heart
Prevent cardiovascular diseases
using Salat Olive Oil
Salat

GRAPHIC Joaquin Brenes, Frank Fernandez   CORPORATION Tribu Saatchi & Saatchi, San José, Costa Rica

ENO
Explore the world

ENO
Explore the world

ROUTINE
SUCKS.
TERRA.COM.BR/TURISMO
terra
travel

ROUTINE
SUCKS.
TERRA.COM.BR/TURISMO
terra
travel

GRAPHIC Gustavo Victorino CORPORATION DDB, São Paulo COPY Otavio Schiavon

NINJA
NINJA
8
90

Miller
Genuine Draft

Pizza Hut
Now serving pasta

Pizza Hut
Now serving pasta

PHOTO Pradeep Dasgupta, Indranie Dasgupta   GRAPHIC Sandeep Salariya, Anirudh Verma, Tanurupa Pal

PHOTO Jonathan Tay   GRAPHIC Amy Koo, Nick Morgan   CORPORATION JWT, Shanghai   COPY Rafael Freire, Morgan

PHOTO Mitch Jenkins

PHOTO Alex ten Napel   GRAPHIC John de Vries

PHOTO Vincent Dixon   GRAPHIC Fernando Riveros, Claudio Campisto, Sergio Araya   CORPORATION BBDO, Santiago de Chile

30
OXFORD
KYG661

Club
STIR
SOMETHING
DEEP INSIDE

Club
STIR
SOMETHING
DEEP INSIDE

GRAPHIC Blair Kimber   CORPORATION JWT, Sydney   COPY Simon Armour

SOMETIMES. YOU DON'T WANT AN ITALIAN MEAL TO TAKE HOURS.
Authentico Pronto
Leggo's
Stir Through
tomato & garlic
with red wine

SOMETIMES, YOU DON'T WANT AN ITALIAN MEAL TO TAKE HOURS.
Authentico Pronto
Leggo's
Stir Through

PHOTO Andreas Smetana   GRAPHIC Tim Holmes   CORPORATION Grey, Melbourne   COPY Brendon Guthrie

PHOTO David Sykes   GRAPHIC Pete Davis   CORPORATION Y&R, Singapore

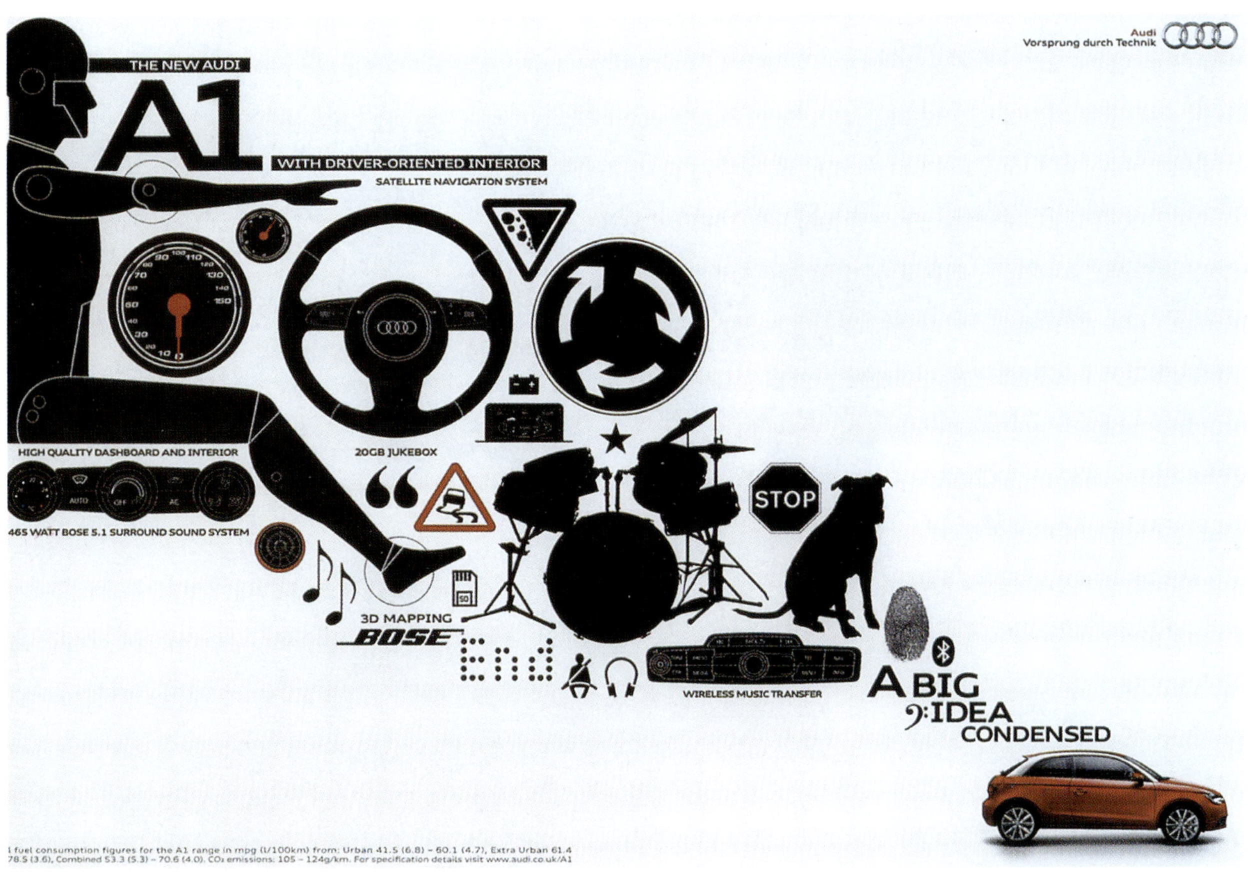

Audi
Vorsprung durch Technik
THE NEW AUDI
A1
WITH DRIVER-ORIENTED INTERIOR
SATELLITE NAVIGATION SYSTEM
HIGH QUALITY DASHBOARD AND INTERIOR
20GB JUKEBOX
465 WATT BOSE 5.1 SURROUND SOUND SYSTEM
3D MAPPING
BOSE
End
STOP
WIRELESS MUSIC TRANSFER
A BIG IDEA CONDENSED
l fuel consumption figures for the A1 range in mpg (l/100km) from: Urban 41.5 (6.8) – 60.1 (4.7), Extra Urban 61.4
78.5 (3.6), Combined 53.3 (5.3) – 70.6 (4.0). CO₂ emissions: 105 – 124g/km. For specification details visit www.audi.co.uk/A1

Audi
Vorsprung durch Technik
HEADLIGHTS INSPIRED BY AN EAGLE'S EYE
THE NEW AUDI
A1
WITH ICONIC EXTERIOR DESIGN
ROOF CONTRAST LINE
POLISHED TWIN EXHAUST
REAR PARK ASSIST
5-TWIN SPOKE ALLOY WHEELS
A BIG IDEA CONDENSED
l fuel consumption figures for the A1 range in mpg (l/100km) from: Urban 41.5 (6.8) – 60.1 (4.7), Extra Urban 61.4
78.5 (3.6), Combined 53.3 (5.3) – 70.6 (4.0). CO₂ emissions: 105 – 124g/km. For specification details visit www.audi.co.uk/A1

GRAPHIC Mark Reddy, Kevin Stark, Nick Kidney, Adrian Rossi   CORPORATION BBH, London   COPY Alex Grieve

PHOTO Kerry Wilson   GRAPHIC AS Anam   CORPORATION Y&R, Beijing   COPY Nils Andersson, AS Anam

At 18.1 kmpl, Maruti Alto lets you explore with its amazing mileage.
ALTO let's go

At 18.1 kmpl, Maruti Alto lets you explore with its amazing mileage.
ALTO let's go

GRAPHIC Harpal Singh   CORPORATION Lowe, Gurgaon   COPY Shayondeep Pal

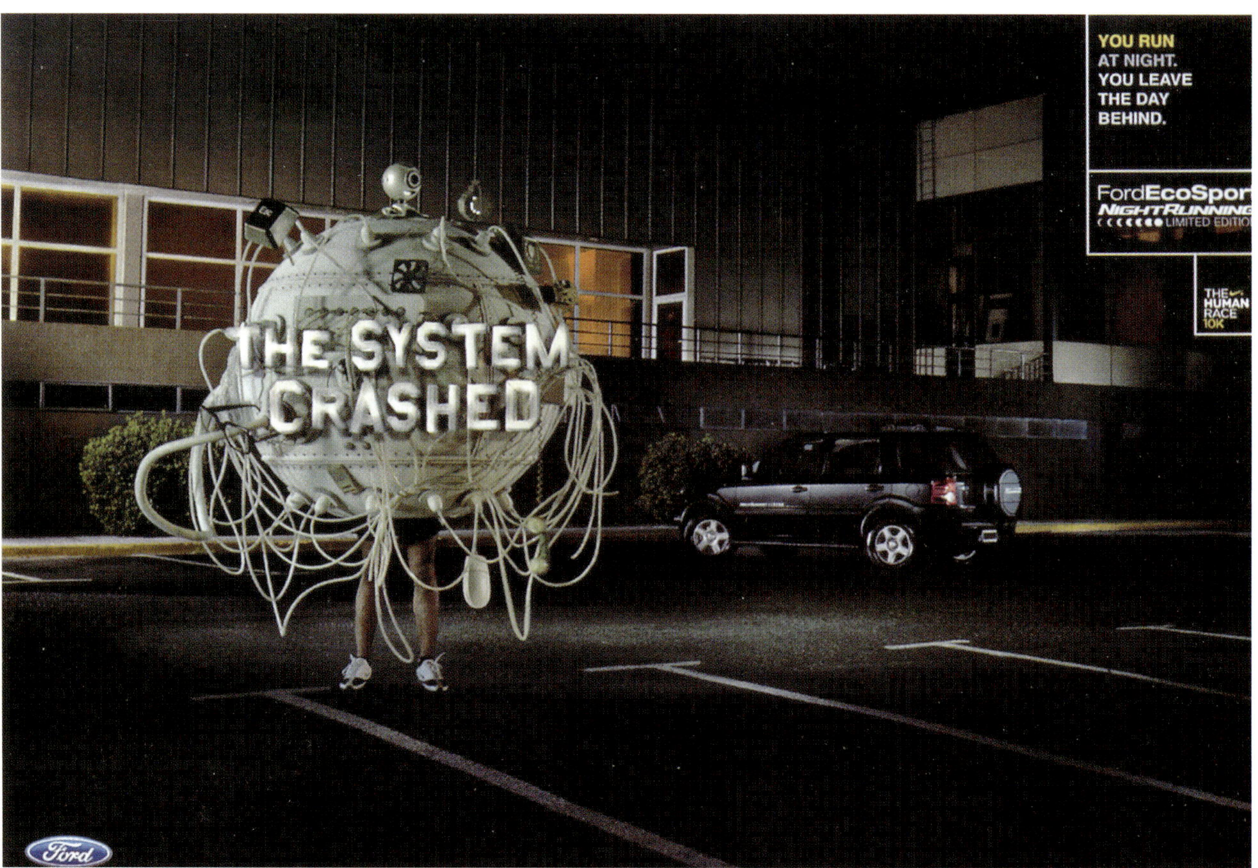

YOU RUN
AT NIGHT.
YOU LEAVE
THE DAY
BEHIND.
Ford EcoSport
NIGHT RUNNING
LIMITED EDITION
THE HUMAN RACE 10K
THE SYSTEM CRASHED
Ford

YOU RUN
AT NIGHT.
YOU LEAVE
THE DAY
BEHIND.
Ford EcoSport
NIGHT RUNNING
LIMITED EDITION
THE HUMAN RACE 10K
4 HOURS FLIGHT
1 HOUR MEETING
4 HOURS FLIGHT
Ford

PHOTO Alejandro Burset   GRAPHIC Miguel Usandivaras   CORPORATION JWT, Buenos Aires   COPY Joaquin Espagnol

A four-door car with a coupé effect.
A5 Sportback.
Audi
Vorsprung durch Technik

Audi
Vorsprung durch Technik
Avek Otomotiv
Roadside assistance at your feet. 0542 443 28 35

PHOTO Ilkay Muratoglu   GRAPHIC Ali Bati   CORPORATION DDB&Co., Istanbul   COPY Gokhan Akca

THE
CHAMBERS OF COMMERCE & INDUSTRY, NORTHERN MIDDLESEX
100TH
ANNIVERSARY
1890
1990

1990

APRIL
PRESS
EDITOR

PHOTO Tom Nagy   GRAPHIC Stefan Schulte, Johannes Hicks, Marc Isken   CORPORATION DDB, Berlin

KTM Made for off-road.

HOW FAR CAN YOU GO?
Jeep

GRAPHIC Ali Bati, Kadir Kaya   CORPORATION DDB&Co., Istanbul   COPY Gokhan Akca

IN A KIA
SPORTAGE ANY
LANDSCAPE
SEEMS FLAT.
KIA
KIA MOTORS
The Power to Surprise

IN A KIA
SPORTAGE ANY
LANDSCAPE
SEEMS FLAT.
KIA
KIA MOTORS
The Power to Surprise

GRAPHIC Juan Jose Posada   CORPORATION Ogilvy & Mather, Bogotá   COPY John Forero

Business Meeting
Cinema
Marina
Kayaking
Jazz Club
Skiing Holiday
MP3 Player
Overnight Bag
Sunglasses
Sea Sickness Pills
Picnic Hamper
Guide Book
Swimming Trunks
Life Jacket
www.landrover.com
In the new Discovery 4 you've the ability to do what you want when you want. Whilst you relax in the super soft leather seats, its new 3.0l diesel engine has the power to make the steepest mountain drive to your ski chalet a breeze - without compromising on economy. And with its intelligent technology you won't even notice you are towing your 26 ft yacht to the marina. So why indeed get a car, when you can get a truly versatile vehicle?
landrover.com/ourplanet
THE NEW 2010 DISCOVERY 4   Why get a car when you can get a Land Rover?
LAND-ROVER
GO BEYOND

www.landrover.com
Swimming Pool
Business Trip
Tables
Fishing Lake
City Centre
Spa
Restaurant
Jewellers
Shoe Shop
Boutique
Keyless Entry
Café
Lingerie Shop
Department Store
The new Discovery 4 looks as good on the inside as it does on the outside. For comfort and ease of use, there are super soft leather seats and improved ergonomics. Also its 5 surround cameras actually make driving safer and easier too. And if you've got your hands full of shopping or ski equipment, no problem, it has keyless entry. So why indeed get a car, when you can get a truly versatile vehicle?
landrover.com/ourplanet
THE NEW 2010 DISCOVERY 4   Why get a car when you can get a Land Rover?
LAND-ROVER
GO BEYOND

PHOTO Andy Glass   GRAPHIC Tim Brookes   CORPORATION RKCR/Y&R, London   COPY Phil Forster

Stop imagining what it smells like and just get rid of it.
SPANJAARD
COCKPIT

Stop imagining what it smells like and just get rid of it.
SPANJAARD
COCKPIT

PHOTO Des Ellis   GRAPHIC Darren Borrino, Greig Watt   CORPORATION DDB, Johannesburg

GRAPHIC Fernando Franco, Jorge Cruz   CORPORATION Maruri Grey, Quito   COPY Fernando Franco, Jorge Cruz

PEPSI

evian
natural spring water
IMPORTED · NON CARBONATED

Extremely quiet.
Touareg TDI® technology.
Das Auto.

Extremely quiet.
Touareg TDI® technology.
Das Auto.

PHOTO Yona Heckel, Pim Vuik   GRAPHIC Ralf Nolting, Christoph Stricker, Eduardo Inderbitzin

Absurdly low consumption.
The Polo BlueMotion.

CLOSED
E
F
Absurdly low consumption.
The Polo BlueMotion.

GRAPHIC Kristoffer Heileman   CORPORATION DDB, Berlin   COPY Ludwig Berndl, Marcus Intek

GRAPHIC Carlos Carrasco

PROTECTING WHO YOU LOVE IS A NATURAL INSTINCT.
IMPREZA XV.
THE CROSSOVER
WITH SUBARU
EXCLUSIVE SAFETY.
SUBARU

PROTECTING WHO YOU LOVE IS A NATURAL INSTINCT.
IMPREZA XV.
THE CROSSOVER
WITH SUBARU
EXCLUSIVE SAFETY.
SUBARU

PHOTO Winkler + Noah   GRAPHIC Marco Turconi   CORPORATION Goettsche, Milan   COPY Elio Buccino

Dr Tint
TINTED WINDOWS
FEEL LIKE A STAR

Dr Tint
TINTED WINDOWS
FEEL LIKE A STAR

PHOTO Andy Anderson   GRAPHIC Jimmy Bonner   CORPORATION The Richards Group, Dallas   COPY Clint Carter

THE ALL-NEW RAM HEAVY DUTY.
MOTOR TREND'S 2010 TRUCK OF THE YEAR.

LONG LIVE THE MAN
WHO STILL LETS
HIS RIGHT FOOT DO
THE TALKING.

RAMTRUCKS.COM

RAM

5.7L V8
HEMI
383 HP
400 LB-FT
OF TORQUE

ALL-NEW
INTERIOR
REFINED

AVAILABLE
INTEGRATED
TRAILER
BRAKE CONTROL

ALL-NEW
CREW
CAB

THE ALL-NEW RAM HEAVY DUTY.
MOTOR TREND'S 2010 TRUCK OF THE YEAR.

PAVE THE WAY IN
PLACES THAT WILL
NEVER SEE THE
LIKES OF PAVEMENT.

RAMTRUCKS.COM

RAM

6.7L CUMMINS
TURBO DIESEL

MASSIVE
BRAKES

ALL-NEW
CREW
CAB

17,600
POUNDS
TOWING
CAPABILITY

PHOTO Andy Anderson   GRAPHIC Jimmy Bonner   CORPORATION The Richards Group, Dallas   COPY Clint Carter

The 4x4 Touareg R50. Powered by 345bhp.

Extremely quiet.
Touareg TDI technology.
Das Auto.

PHOTO Dick Chan   GRAPHIC O'Poon, Fei Leung   CORPORATION DDB, Hong Kong   COPY Paul Chan

A Ford travels by train
Our cars are transported from the factory by train. This saves resources and helps the environment.
www.ford.dk/environment
Feel the difference
Ford

A Ford wears old jeans
The old jeans are used for noise reducing material. This saves resources and helps the environment.
www.ford.dk/environment
Feel the difference
Ford

PHOTO Søren Hald   GRAPHIC Claus Collstrup   CORPORATION Ogilvy & Mather, Copenhagen   COPY Mikkel Elung

Eine Marke der Daimler AG

Benz 16 mit der Abstandsregelung DISTRONIC PLUS, 1926.
Mercedes-Benz – stets um Innovation bemüht.

Mercedes-Benz

Eine Marke der Daimler AG

508-M

Mercedes Simplex 508 mit Intelligent Light System, 1902.
Mercedes-Benz – stets um Innovation bemüht.

Mercedes-Benz

GRAPHIC Volkmar Weiss, Stefan Mayer   CORPORATION Jung von Matt, Vienna

KIA SPORTAGE,
now with GPS.
KIA
KIA MOTORS
The Power to Surprise
AV CENTRAL CALL
EVA

KIA SPORTAGE,
now with GPS.
KIA
KIA MOTORS
The Power to Surprise
CAL R
V RIO AV CA

GRAPHIC Juan Posada   CORPORATION Ogilvy & Mather, Bogotá   COPY Diego Ortiz, John Forero

P
WASH ME!
Imagine taking as
much shit as your car.
We have everything you need to take good care of your car.
Welcome to our petrol stations.
OK Q8

smart
open your mind.
>> new smart with electronic stability programme.
AVENIR
Nuevo Toyota
www.toyota.es

PHOTO Petrus Olsson   GRAPHIC Richard Baynham   CORPORATION Ogilvy & Mather, Stockholm   COPY Per-Olov Lundgren

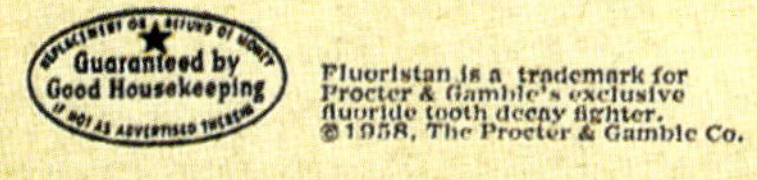

RIDGEWOOD DENTAL CLINIC
To the parents of: Johnny Bolton
I have examined Johnny's teeth
and find no new cavities
NBN D.D.S.

Norman
Rockwell

"Look, Mom — no cavities!"

Crest Toothpaste with fluoride means far fewer cavities for every member of the family—including children of all ages.

Crest's fluoride is the same decay fighter
that dentists apply directly on teeth.
Ask your dentist about Crest.

Crest
TOOTH PASTE
with fluoristan

Guaranteed by
Good Housekeeping

Fluoristan is a trademark for
Procter & Gamble's exclusive
fluoride tooth decay fighter.
©1958, The Procter & Gamble Co.

CORPORATION Leo Burnett, Chicago

Lite
A FINE PILSNER

THE SPECTACULAR
TRAIN
EXHIBITION
19 NOVEMBER '89
1 JANUARY '90
EMPORIUM
SAN FRANCISCO
CENTRE

Extravagant.
Exorbitant.
Extortionate.
Extremely reasonable.
STELLA ARTOIS
Stella is brewed with fragrant female Saaz hops, found only on the banks of the Eger. (They don't grow on trees, we can tell you.)
And an unadulterated mix of the world's finest barleys. (At the world's highest prices.)
Having transmuted these into liquid gold, we sit on our assets for months, while the brew matures to perfection, and our accountants are driven to distraction.
So it's not surprising people gasp in astonishment on hearing the price of a round of Stella. It's far too cheap.
Stella Artois. Reassuringly expensive.

YUGOSLAVIA
BELGIUM
CZECHOSLOVAKIA
HOLLAND
GERMANY
FRANCE
Who pays for our annual trip round Europe? You'll find out when it's your round.
You might have wondered, whilst paying for a round of Stella, where all that money goes.
Cruises up the Eger gathering the fragrant Saaz hops?
Trips through the choicest barley fields of the world?
It's true, we spend some money in these areas. But a vast amount goes on sitting around doing nothing.
However long it takes a particular brew of Stella to mature we just have to wait, and wait.
Six weeks, two months or even longer, we can't sell a drop till our head brewer says the moment is right.
It plays havoc with our cash flow.
Almost as much as drinking Stella does with yours.
Stella Artois. Reassuringly expensive.

PHOTO Graham Ford   GRAPHIC Dave Christensen   CORPORATION Lowe, London   COPY Patrick Woodward

Birgit & Berlin
Example: Beautiful used Tarpaulin
KULTBVG
Great One-Of-Kinds Are Drawn To The Same.
KULTBVG
Beautiful used since 1996

Wilma & Walhalla
Example: Beautiful used Tarpaulin
KULTBVG
Great One-Of-Kinds Are Drawn To The Same.
KULTBVG
Beautiful used since 1996

Manuela & MalibuBeach
Example: Beautiful used Air-Mattress
KULTBVG
Great One-Of-Kinds Are Drawn To The Same.
KULTBVG
Beautiful used since 1996

Simone & Shopper
Example: Beautiful used Tarpaulin
KULTBVG
Great One-Of-Kinds Are Drawn To The Same.
KULTBVG
Beautiful used since 1996

PHOTO Nannette Römer   GRAPHIC Markus Kreykenbohm

THIS IS SO SUDDEN
T. FLUHARTY '92

FLUHARTY '92

QUILL BROS.
SINCE 1707
T. FLUHARTY

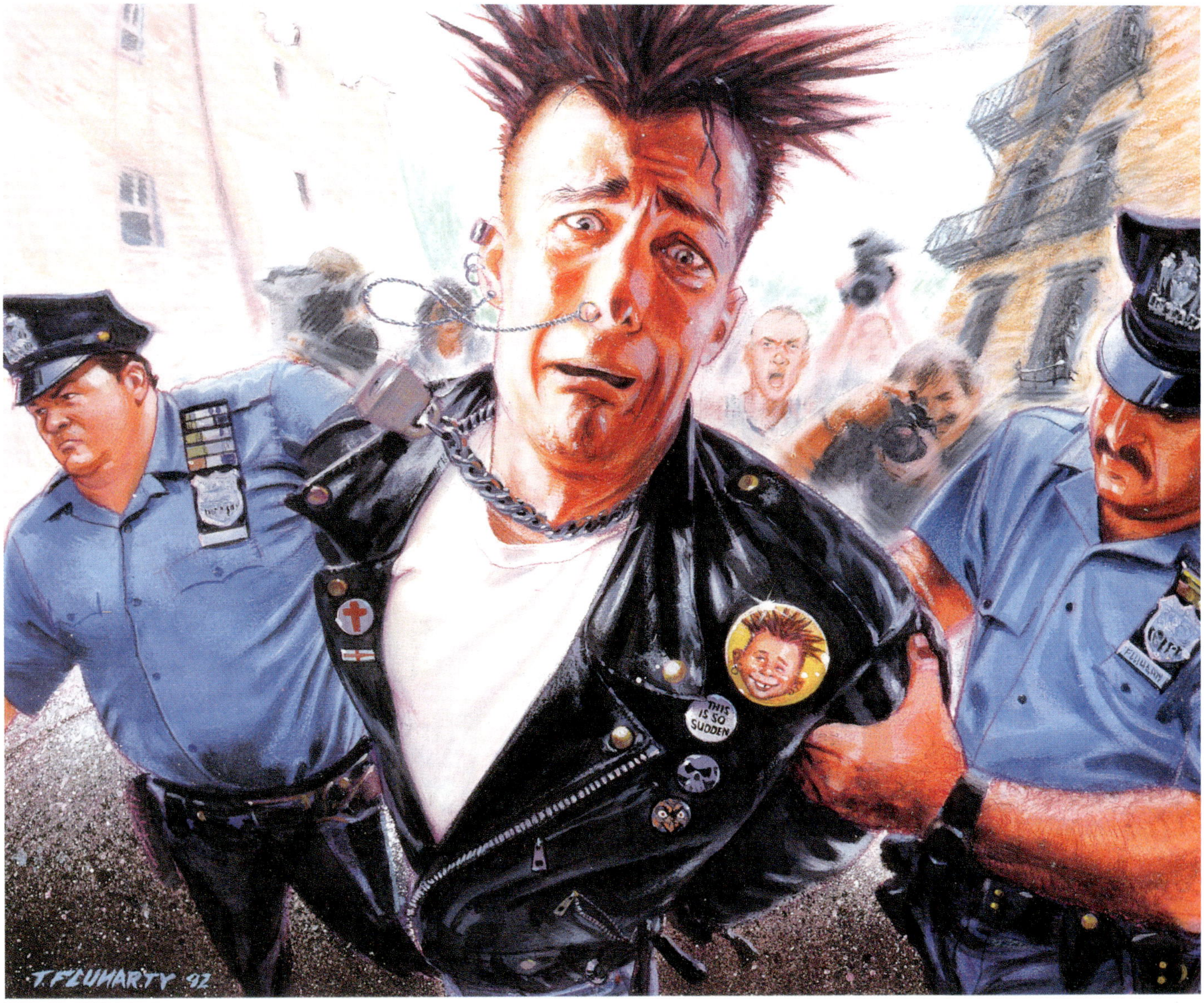

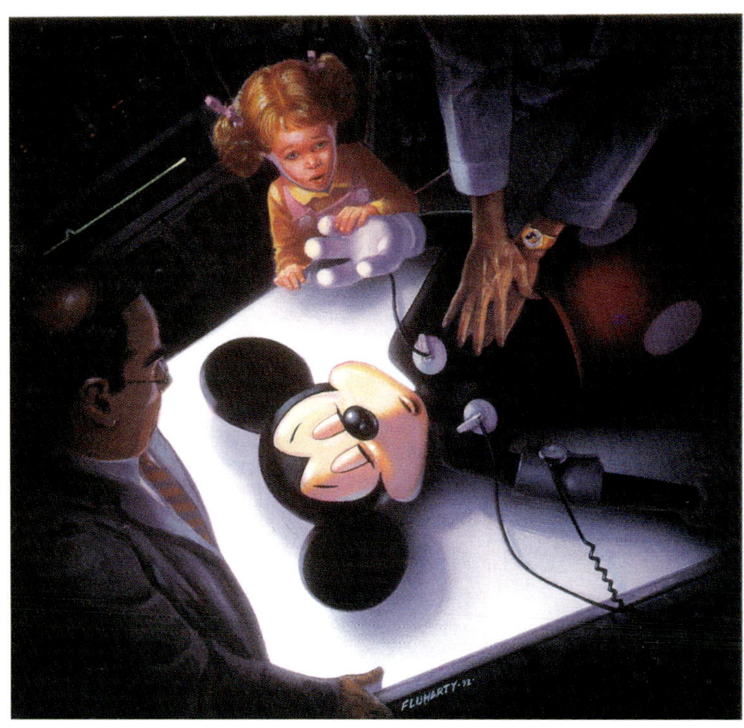

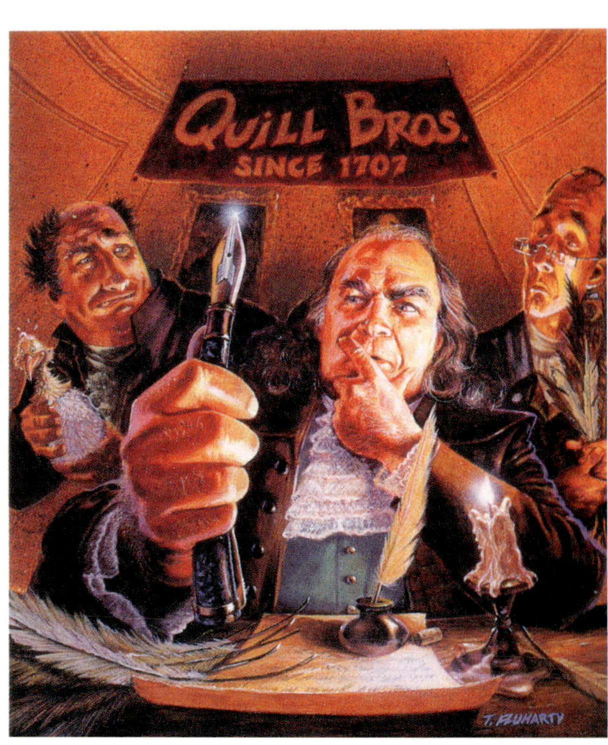

GRAPHIC Adam Livesey   CORPORATION TBWA\Hunt\Lascaris, Johannesburg   COPY Matthew Brink

AJAX
Never-ending shine.

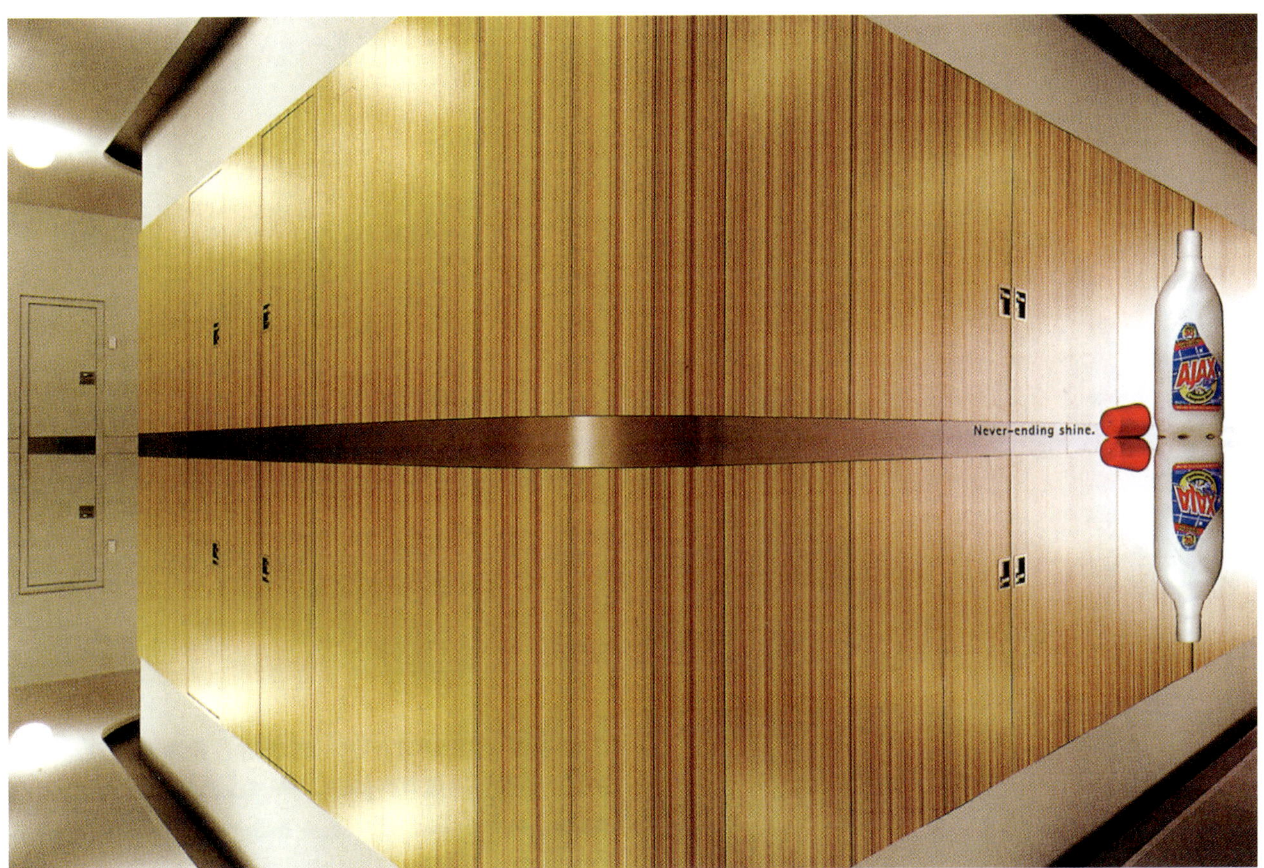
AJAX
Never-ending shine.

PHOTO Fernando Zuffo   GRAPHIC Marco Mattos   CORPORATION Y&R, São Paulo   COPY Vinicius Stanzione

Don´t let your colours mix.
Tide

Don´t let your colours mix.
Tide

INTENSITY
THE SCENT OF REAL CLEAN
Pine-Sol

INTENSITY
THE SCENT OF REAL CLEAN
LEMON FRESH
Pine-Sol

Pine-Sol
INTENSITY
THE SCENT OF REAL CLEAN

PHOTO Matthew Rolston   GRAPHIC Peter Barnes   CORPORATION DDB West, San Francisco   COPY Lisa Goodfriend

CIENT FOREST CRIME
AMAZON
SA
Low VOC paint
Metalatex Eco.
Helps to protect
the environment.

Low VOC paint
Metalatex Eco.
Helps to protect
the environment.

PHOTO Mário Daloia   GRAPHIC Marcus Kawamura   CORPORATION Mohallem Meirelles, São Paulo   COPY Ana Carolina Reis

GLUES ANYTHING.
AS LONG AS IT'S PAPER.
Pritt

GLUES ANYTHING.
AS LONG AS IT'S PAPER.
Pritt

GLUES ANYTHING.
AS LONG AS IT'S PAPER.
Pritt

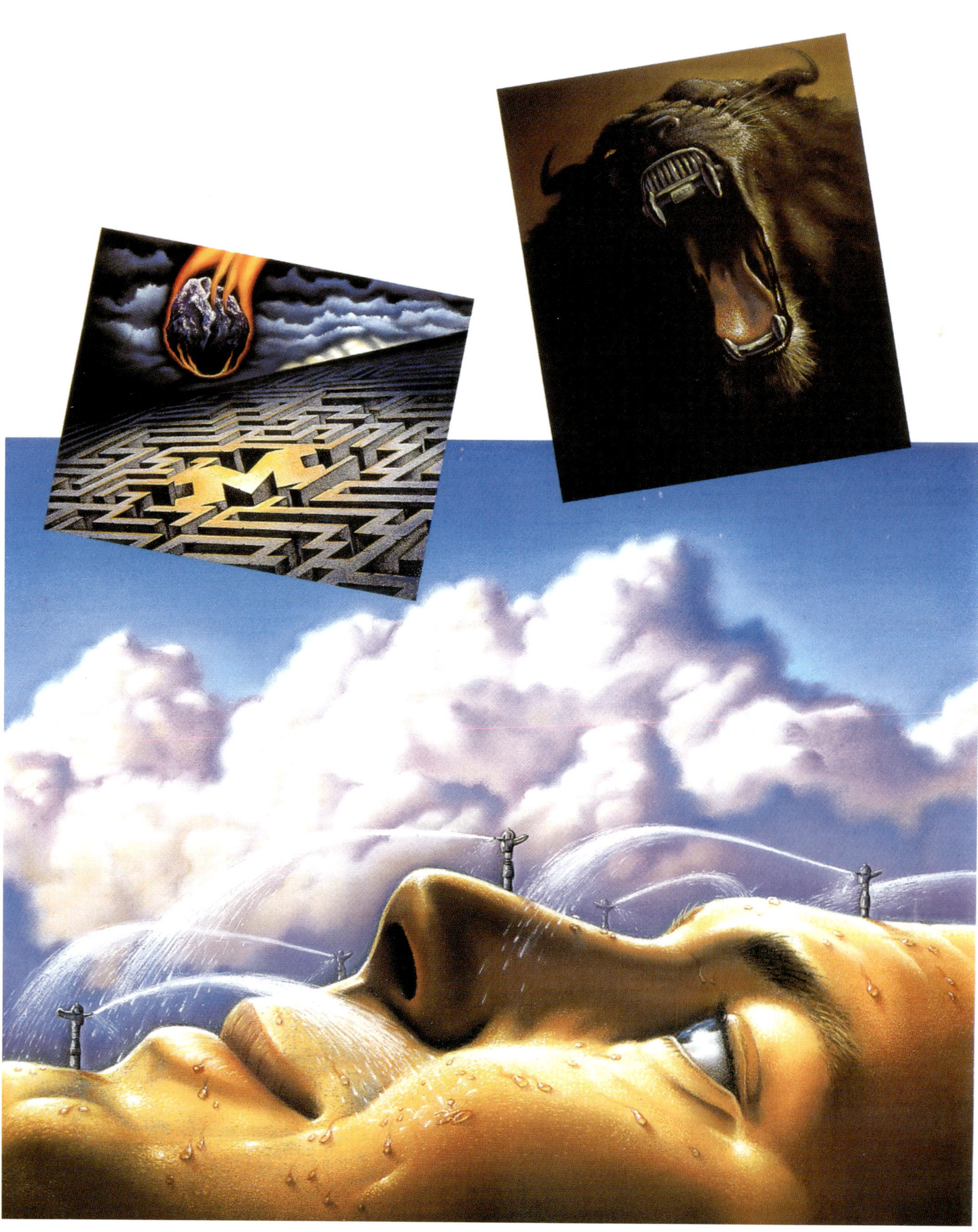

The Girl Who Loved Caterpillars
ADAPTED BY JEAN MERRILL · ILLUSTRATED BY FLOYD COOPER

EXTENDS LIFE
Sekunda. Universal glue

EXTENDS LIFE
Sekunda. Universal glue

PHOTO York Christoph   GRAPHIC Björn von Buchholtz, Zuzana Havelcová, Timm Holm   CORPORATION Butter, Berlin

GENERAL
COMPARTMENT

FOR A PERFECT WHITE.
Swing Color
ARCTIC WHITE

febreze
GET YOUR
HOME BACK

GRAPHIC Cedric Auzannet   CORPORATION Grey, Paris   COPY Benjamin Dessagne

THE WONDERFUL WORLD OF Vh1

THE WONDERFUL WORLD OF Vh1
DAD
BROTHER
MOM
SISTER
MUDDY

GRAPHIC Alexander Nowak   CORPORATION Y&R, New York   COPY Feliks Richter

Ist es eine Sünde, wenn man nicht alle E-Mails beantwortet?
Sind wir reif für eine neue Zeitung?
KURZ. ANDERS.
GEDRUCKT.
WELT KOMPAKT
Die Rückkehr der künstlichen Welten im Internet
sind-wir-reif.de

Wir melden unsere Lieblinge bei Facebook an.
Sind wir reif für eine neue Zeitung?
KURZ. ANDERS.
GEDRUCKT.
WELT KOMPAKT
Die Rückkehr der künstlichen Welten im Internet
sind-wir-reif.de

PHOTO Sascha Weidner   GRAPHIC Torsten Lesszinsky   CORPORATION Dorten, Stuttgart   COPY Christian Schwarm

SKY
Peace on earth and war up in the skies.

SKY
All against all. And viceversa.

GRAPHIC Paolo Perrone, Giorgio Cignoni   CORPORATION 1861united, Milan   COPY Federico Ghiso

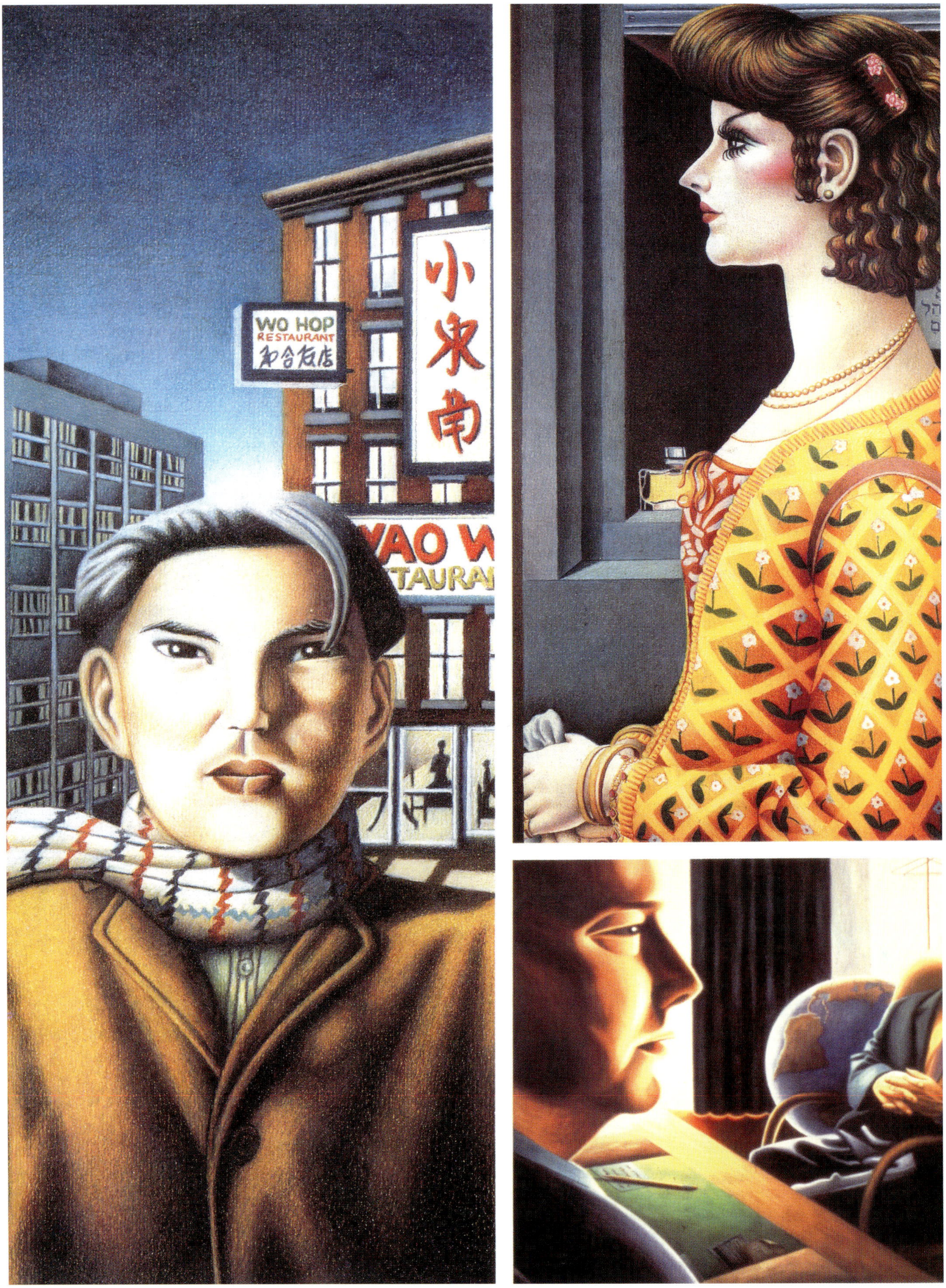

FEDERAL EXPRESS
Overnight
#1

Mobil Masterpiece Theatre presents
Samuel Richardson's
Clarissa

The virtuous victim just wouldn't submit.

A three-part series begins Sunday, April 5   9 PM on PBS

With Saskia Wickham as Clarissa and Sean Bean as Lovelace

Host: Alistair Cooke

194 THIRD AVENUE NEW YORK NY 10003
VICKI MORGAN ASSOCIATES
(212) 475·0440

BEAUTY IDEALS CHANGE.
HELP THE VICTIMS TO SURVIVE:
WWW.ANTIANOREXIA.NET
ANAD

BEAUTY IDEALS CHANGE.
HELP THE VICTIMS TO SURVIVE:
WWW.ANTIANOREXIA.NET
ANAD

GRAPHIC Albert S. Chan, Christian Mommertz   CORPORATION Ogilvy & Mather, Frankfurt am Main

Your beauty. Up in smoke.
Tobacco will destroy your face cell by cell, from the inside out. That's the ugly truth.
WORLD NO TOBACCO DAY
McCANN HEALTHCARE WORLDWIDE
Ugly Truth

Your future together. Up in smoke.
Secondhand smoke will steal her looks, her youth, and her health. That's the ugly truth.
WORLD NO TOBACCO DAY
McCANN HEALTHCARE WORLDWIDE
Ugly Truth

Your health. Up in smoke.
Ugly Truth

PHOTO Tomiyuki Takahashi   GRAPHIC Hajime Nakazawa, Misuzu Inoue

Fordelene
ved å være mange
er mange
Med 865.000 medlemmer får
vi til svært gode avtaler for deg,
også i privatlivet. Les mer
om fordelene på lofavor.no
lo favor

FORDELENE
VED Å VÆRE MANGE
ER MANGE
Med 865.000 medlemmer får
vi til svært gode avtaler for deg,
også i privatlivet. Les mer
om fordelene på lofavor.no
lo favor

PHOTO Petrus Olsson   GRAPHIC SMFB, Oslo   CORPORATION Carl-Erik Conforto   COPY Stig Bjølbakk

S H K D
SAHIPSIZ HAYVANLARI
KORUMA DERNEĞI
Everyone deserves a second chance.
Pay a visit to our shelters and
make one of our dogs a "lucky" pet.

Everyone deserves a second chance.
Pay a visit to our shelters and
make one of our dogs a "lucky" pet.
S H K D
SAHIPSIZ HAYVANLARI
KORUMA DERNEĞI

PHOTO Rainer Stratmann   GRAPHIC Boran Erem   CORPORATION BBDO, Istanbul   COPY Arda Albayraktar

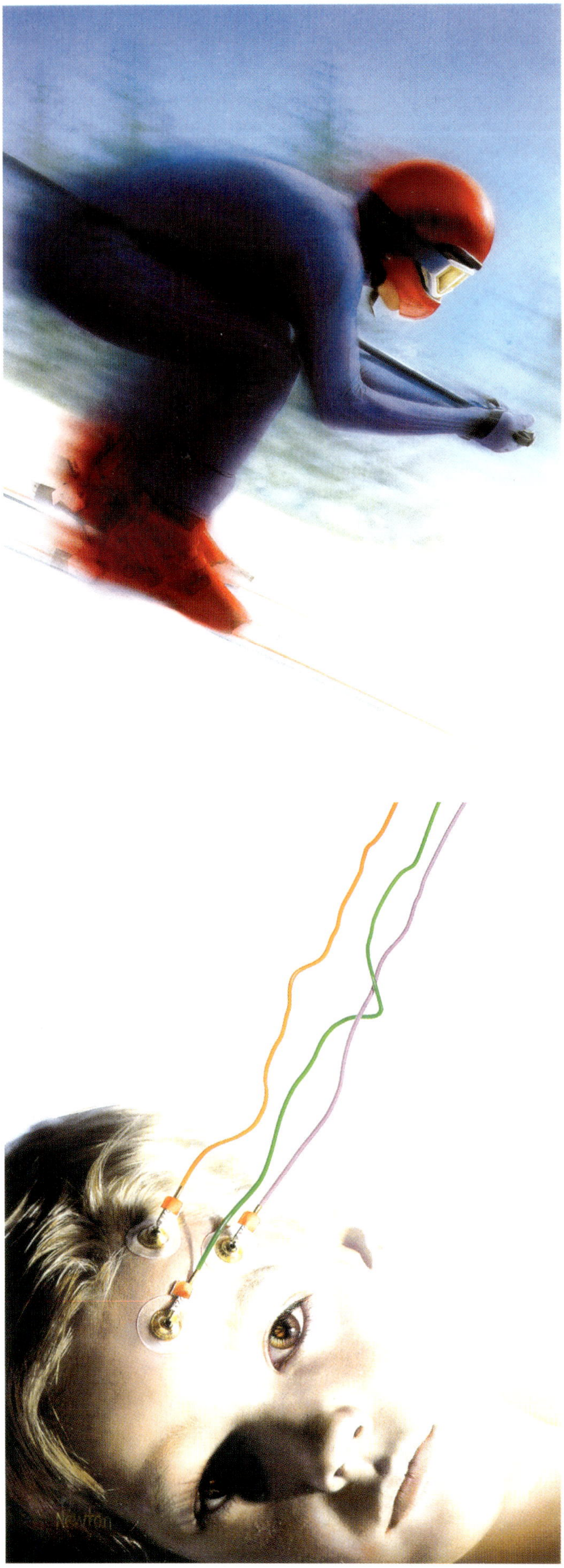

Newton

AUTO
10M [IN 25]
Nikon
The Nikon S60. Detects up to 12 faces.

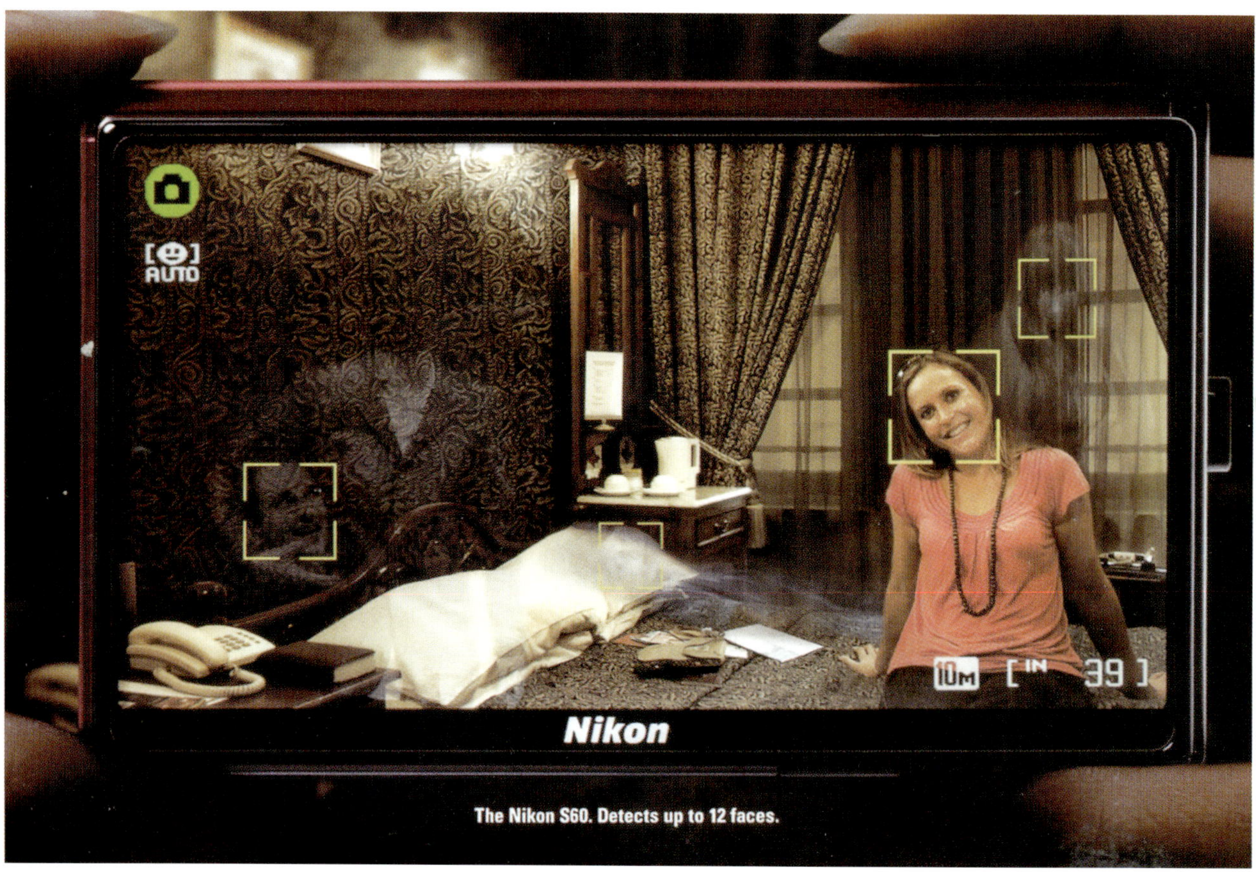
AUTO
10M [IN 39]
Nikon
The Nikon S60. Detects up to 12 faces.

PHOTO Chai Guan Teo   GRAPHIC Somjai Satjatham   CORPORATION Y&R, Singapore   COPY Y&R, Singapore

mint
museum of toys
SOME OF THE FIRST CARS JAPAN EXPORTED
WERE NOTHING MORE THAN TIN CANS.
THEY WERE RIGHT.
MINT MUSEUM OF TOYS
26 SEAH STREET
OPEN DAILY 9.30AM - 6.30PM
WWW.EMINT.COM

The gospel. Truth?
St. Matthew-in-the-City

FISHY
CATERING
The gospel. Truth?
St. Matthew-in-the-City

AUTO
Nikon
10M [1N 6 1
The Nikon S60. Detects up to 12 faces.

OLYMPUS
E-3
Nothing Escapes. E-3 World's Fastest Auto Focus.
OLYMPUS

PHOTO Kwok Tim Lau   GRAPHIC Wendy Chan, Zhong Qing Liu, Zhi Guang Ye   CORPORATION Saatchi & Saatchi, Guangzhou

SONY
make.believe
Believe in capturing incredible
moments in low light.
Iguazu Falls, Argentina, 17:39

SONY
make.believe
Believe in capturing incredible
moments in low light.
Iguazu Falls, Argentina, 17:54

PHOTO Pedro Alvarez   GRAPHIC Ben Beazley, Kyla   CORPORATION Fallon, London   COPY Ben Beazley, Kyla Elliott

Pay
a visit
back
to your
past
ISTANBUL
TOY
MUSEUM

PAY
A VISIT BACK
TO YOUR
PAST.
ISTANBUL
TOY
MUSEUM

PHOTO Suleyman Kacar GRAPHIC Selim Ünlüsoy CORPORATION Markom/Leo Burnett, Istanbul COPY Bahadirhan Peksen

Life doesn't stop for TV
Record or pause all your favourite docos
FOXTELiQ

SPORTS HD
QANTAS
HYUNDAI
GAME ON
SOLO
FOX SP
Life doesn't stop for TV
Record or pause all your favourite matches
FOXTELiQ

PHOTO Neil Bailey   GRAPHIC Noah Regan, Becky Alperstein   CORPORATION Three Drunk Monkeys, Sydney

KRAFT
Miracel
Whip
so leicht
4,9
Only 4,9% Fat.

PHOTO Adrian Chan, Maurice Wee, Stuart Mills   GRAPHIC Jula Böhm   CORPORATION Draftfcb, Hamburg

There's a much
Juicier
chew
Juicy Fruit

VEGETABLES
turnip
(brassica campestris)
spanish onion
(allium cepa)
red onion
asparagus
(officinalis)
shalot
garlic (allium sativum)
red pepper
leek
(porum)
green pepper
(capsicum)
eggplant
(solanum melones)
carrot
(daucus carrota)

Close Encounters.
Cincinnati Zoo.

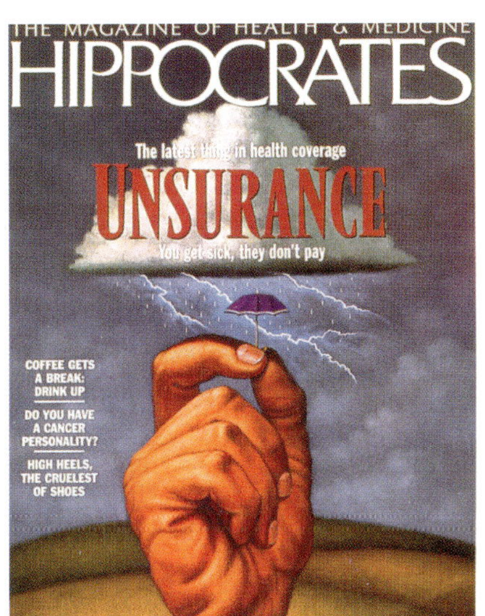

THE MAGAZINE OF HEALTH & MEDICINE
HIPPOCRATES
The latest thing in health coverage
UNSURANCE
You get sick, they don't pay
COFFEE GETS
A BREAK.
DRINK UP
DO YOU HAVE
A CANCER
PERSONALITY?
HIGH HEELS,
THE CRUELEST
OF SHOES

Once There Was A Farm...
A Country Childhood Remembered
Virginia Bell Dabney

SNOW 37381 IMAGES
A STORM
www.otherimages.com
OTHER IMAGES
ONE BANK. MILLION IMAGES.

TREE 65648 IMAGES
A FOREST
www.otherimages.com
OTHER IMAGES
ONE BANK. MILLION IMAGES.

This is
0,000044%*
of all our images
gettyimages

Você tem 150 mil fios de cabelos. Veja como organizar tudo
FashionHair
A revista das tendências

Dicas sobre cabelo que você não encontra nem no Google
FashionHair
A revista das tendências

FashionHair
A revista das
tendências.

FashionHair

PHOTO Marcio Rodrigues   GRAPHIC Waldemar Franca

Flowers mAke A bEtter Day
Roßkopf
Gardeners since 1936

Hi
Flowers make A BETTeR dAy
Roßkopf
Gardeners since 1936

GRAPHIC Diana Sukopp, Volker Schrader   CORPORATION Publicis, Frankfurt am Main   COPY Stephan Ganser

„In ihren fruchtbaren Tagen
bevorzugen Frauen stürmische Männer.
In den unfruchtbaren sanfte.
Ein kluger Mann fährt nicht in
beiden Phasen das gleiche Auto."
SiXT
www.sixt.de

„Nur primitive Männer fahren tolle Autos,
um eine Frau rumzukriegen.
Ich mag primitive Männer."
SiXT
www.sixt.de

PHOTO Lado Alexi   GRAPHIC Vanessa Rabea Schrooten, Felix Taubert, Dirk Haeusermann   CORPORATION Jung von Matt, Hamburg

OUR BEAUTY PRODUCTS CAN AFFECT THE FACES OF PEOPLE THOUSANDS OF MILES AWAY.
We always strive to bring you products bursting with effectiveness to enhance your natural beauty. By choosing products made with Community Trade ingredients you are giving people in communities around the world the chance to build a better future. We work directly with our Community Trade suppliers to build businesses in more than 20 countries helping 25,000 people earn a sustainable income. We don't do this because it's fashionable; we do it because we believe it's the only way to do business.
THE BODY SHOP

UNTIL SOMEONE DISCOVERS ANOTHER PLANET WHERE NATURALLY EFFECTIVE INGREDIENTS GROW, WE'LL HELP LOOK AFTER THIS ONE.
We're constantly searching for nature's unique ingredients from all four corners of the world and bringing you products bursting with effectiveness. Whilst we're doing this, we're always looking for new ways to protect this beautiful planet. Our Community Trade organic olive oil, for example, is grown without the use of chemical pesticides. We're also limiting our packaging and introducing bottles made from 100% recycled plastic. So whilst we're helping you look beautiful, we're helping the planet to stay beautiful too.
THE BODY SHOP

GRAPHIC Paul Copeland   CORPORATION Leagas Delaney, London   COPY Rob Burleigh

TAKE THE ROAD
LESS FRAZZLED.
NORTHEAST REGIONAL
AMTRAK
AMTRAK.COM

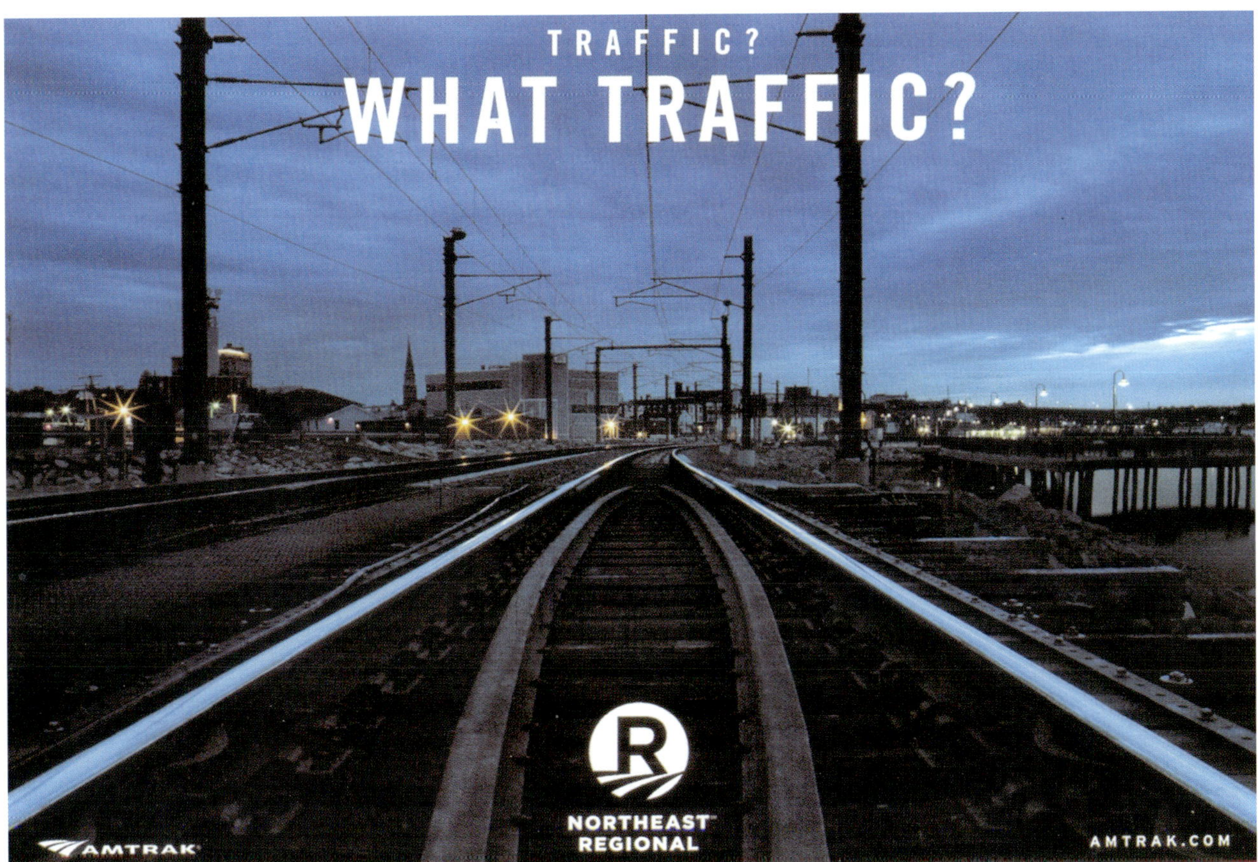

TRAFFIC?
WHAT TRAFFIC?
NORTHEAST REGIONAL
AMTRAK
AMTRAK.COM

PHOTO Jim Smithson   GRAPHIC Tracy Nenna   CORPORATION Arnold, Washington   COPY Trevor Sloan, Lisa Biskin

GRAPHIC Tim Stübane, Stefan Schulte, Gabriel Mattar

PHOTO Erika Lidén, Linda Åkerberg

Fresh All Summer
CALGARY FARMERS' MARKET

Fresh All Summer
CALGARY FARMERS' MARKET

Fresh All Summer
CALGARY FARMERS' MARKET

Fresh All Summer
CALGARY FARMERS' MARKET

PHOTO Ken Woo   GRAPHIC Brad Connell

HELP
DO IT
BE
I have nothing to say.
MIND STOPPER
THIS IS MY CLONE
boring
live love learn
DO YOU H
this bracelet ma
WIRE HANG
NO WIRE H
Jessica Kagan Cushman
Blended style
blender

Christian Louboutin
Blended fashion
blender

FOR PEOPLE WHO LIKE TO SAVE EVERY SEN
QC CONVENIENCE STORE

PHOTO Ho Ming Yeo

FOR PEOPLE WHO LIKE TO SAVE EVERY SEN
QC CONVENIENCE STORE

PHOTO Jesse Choo   GRAPHIC Phoecus Lee, David Sin   CORPORATION Grey, Kuala Lumpur   COPY Gautam Mehta

A few exclamation marks to finish off your fashion statement.
John Lewis
johnlewis.com

We're proud to introduce the eight-piece suit.
John Lewis
johnlewis.com

PHOTO Metz & Racine, John Akehurts   GRAPHIC Emer Stamp   CORPORATION Adam&Eve, London   COPY Ben Tollett

Otrivin
pleasure
of breathing

Otrivin
pleasure
of breathing

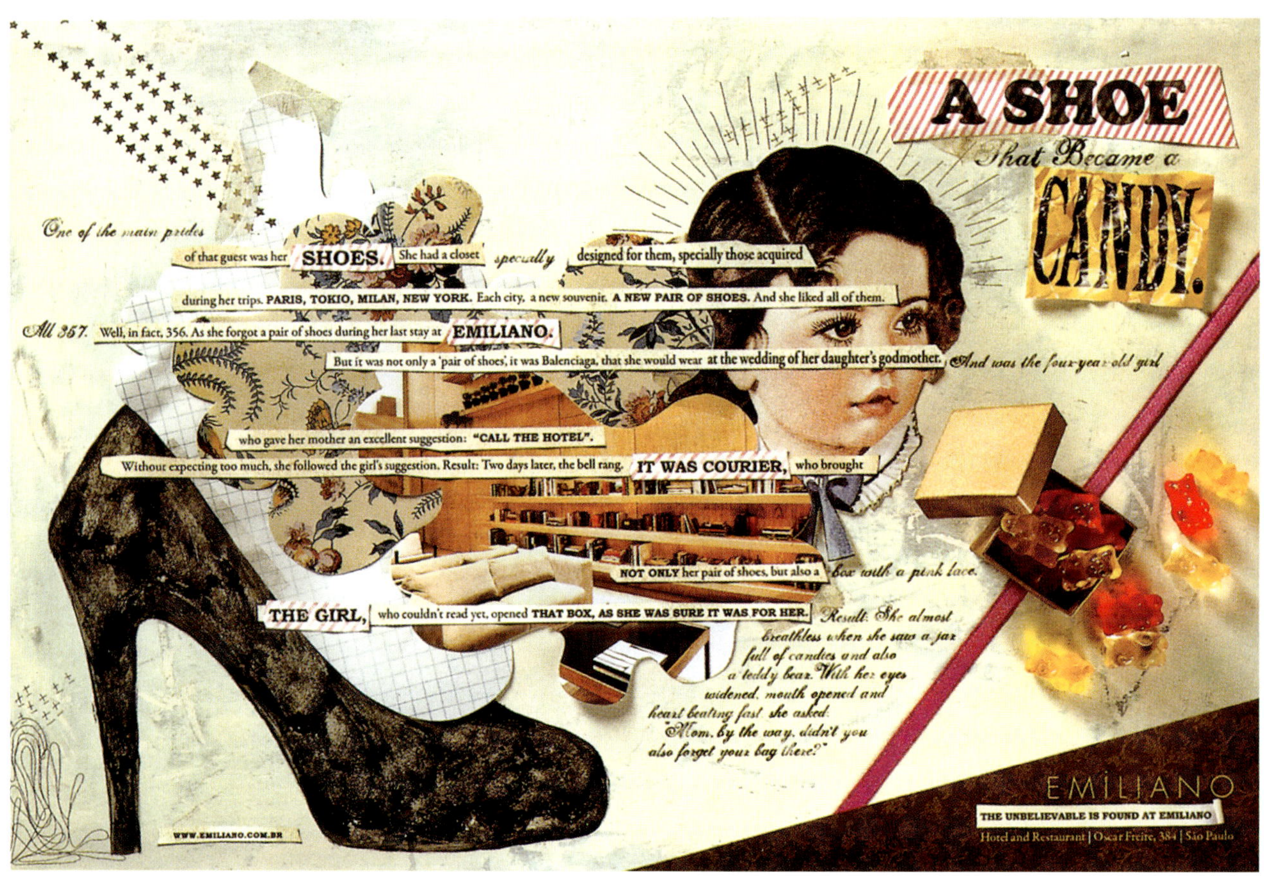
A SHOE
That Became a
CANDY.
One of the main prides
of that guest was her SHOES. She had a closet specially designed for them, specially those acquired
during her trips. PARIS, TOKIO, MILAN, NEW YORK. Each city, a new souvenir. A NEW PAIR OF SHOES. And she liked all of them.
All 357. Well, in fact, 356. As she forgot a pair of shoes during her last stay at EMILIANO.
But it was not only a 'pair of shoes', it was Balenciaga, that she would wear at the wedding of her daughter's godmother. And was the four-year-old girl
who gave her mother an excellent suggestion: "CALL THE HOTEL".
Without expecting too much, she followed the girl's suggestion. Result: Two days later, the bell rang. IT WAS COURIER, who brought
NOT ONLY her pair of shoes, but also a box with a pink lace.
THE GIRL, who couldn't read yet, opened THAT BOX, AS SHE WAS SURE IT WAS FOR HER. Result: She almost
breathless when she saw a jar
full of candies and also
a teddy bear. With her eyes
widened, mouth opened and
heart beating fast, she asked:
"Mom, by the way, didn't you
also forget your bag there?"
EMILIANO
THE UNBELIEVABLE IS FOUND AT EMILIANO
Hotel and Restaurant | Oscar Freire, 384 | São Paulo
WWW.EMILIANO.COM.BR

"Smile of the Dolls"
She was one of the main executives of a multinational
telecommunications company. She frequently traveled on business,
but always took her daughter with her and, consequently, the nanny.
It wasn't a perk, it was need. Her agenda was filled with meetings,
with little time left for the girl. Any help she could get was welcome and necessary.
Especially on that day - Julia and Cecilia's birthday.
However, before leaving for her work, she asked Emiliano's staff
to prepare a party. She wanted them to take care of everything.
"Carte blanche!", she said in French with a Parisian accent.
"Oui!", the hotel staff acknowledged.
They prepared a table with cakes, chocolate truffles, balloons and party gifts.
And, at 6:30 p.m., they made a surprise visit to room 56.
They, however, did not notice a detail: why did the cake only have one candle
if the girl was at least six and the nanny was clearly over forty?
Almost as if she could read their minds,
the girl picked up her two dolls and said:
"This is Julia and this is Cecilia. They're both Leos".
And under a shy expression from the nanny and a sincere smile from the girl,
the staff had no choice but to sing Happy Birthday,
for the delight of the dolls who gave a smile.
At least the girl though they did.
EMILIANO
THE UNBELIEVABLE IS FOUND AT EMILIANO
Hotel and Restaurant | Oscar Freire, 384 | São Paulo
www.emiliano.com.br

GRAPHIC Sthefan Ko   CORPORATION JWT, São Paulo   COPY Luiz Filipin